D0122955

Barry Parker

Einstein

The Passions of a Scientist

Barry Parker

Einstein

The Passions of a Scientist

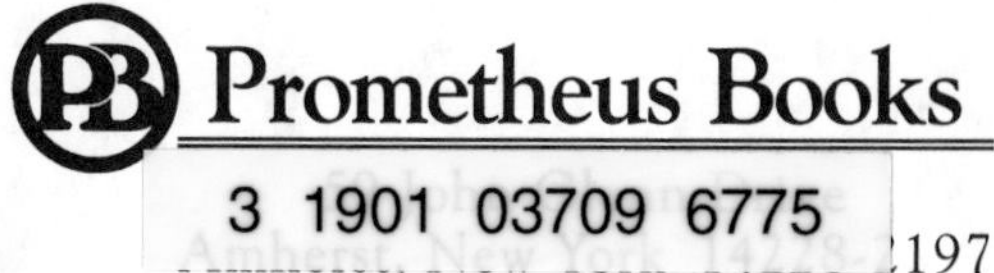

Published 2003 by Prometheus Books

Inquiries should be addressed to
Prometheus Books
59 John Glenn Drive
Amherst, New York 14228–2197
VOICE: 716–691–0133, ext. 207
FAX: 716–564–2711
WWW.PROMETHEUSBOOKS.COM

07 06 05 04 03   5 4 3 2 1

Library of Congress Cataloging-in-Publication Data

Parker, Barry R.
Einstein : the passions of a scientist / Barry Parker.
p. cm.
Includes bibliographical references and index.
ISBN 1–59102–063–8 (alk. paper)
1. Einstein, Albert, 1879–1955. 2. Physicists—Biography. 3. Relativity (Physics)
I. Title.

QC16.E5P367 2003
530'.092—dc21
[B]   2002036942

Printed in Canada on acid-free paper

Contents

Preface

Albert Einstein has now been dead for almost fifty years, yet he still generates a tremendous amount of public interest. And his discoveries are still making news. His "cosmological constant," for example (an attachment to his equations of general relativity, which he cast off in 1932 with the statement "it was the greatest blunder of my life"), is now back in vogue. It predicts a slight speeding up, or acceleration, of the outermost galaxies in the universe, something that seems to go against common sense. Yet recently it has been discovered that these galaxies are, indeed, accelerating.

In 2000 I wrote a book titled *Einstein's Brainchild* in which I focused on Einstein's contributions to science. The present volume is meant to complement it in that it focuses on Einstein's life. Einstein's passions are the central theme of the book, but I also emphasize the interrelationships between them and how they affect one another. His relationships with women also certainly affected his development and work, and I also discuss this.

Einstein was a complex person. He had strong feeling and an ambition to succeed even when he was quite young. Although his family was strongly supportive, there was still considerable turmoil in his life. He was rebellious at school and later at university, and it caused him many problems. Even when he became famous, there were difficulties. We will look at all of them in this book.

A large number of excellent biographies of Einstein have been written in the last few years, and I do not want to try to compete with them. This is not a biography, but rather a detailed look at his obsessions and passions, what effect they had on his development as a scientist, and how they may have helped lead to his breakthroughs. Physics played a large role in Einstein's early life, but he had a life aside from physics that certainly had some bearing on his development as a scientist.

Although I have taken advantage of the extensive literature that is available on Einstein, I have concentrated on the original sources. Upon his death, Einstein left all his papers and other documents to the Hebrew University of Jerusalem. The collection was extensive, containing more than 43,000 items. Over the past few years, copies of almost everything have been collected by Princeton University, and to some degree by Boston University. There are now eight volumes of the collected works of Einstein at Princeton University, and all are available.

It is not possible to write the story of a scientist without using scientific terms. I have tried to explain them as I use them, but for the benefit of those new to science, I have included a glossary at the end of the book.

The drawings were done by Lori Scoffield-Beer based on photographs from the Albert Einstein Archives. I would like to thank her for an excellent job and the Hebrew University of Jerusalem for allowing us to use the photographs. Other photographs used are courtesy of the Lotte Jacobi Archives. I would also like to thank my editor, Linda Greenspan Regan, and the staff of Prometheus Books for their help in bringing this book to its final form. Finally, I would like to thank my wife for her support while it was being written.

Introduction

Albert Einstein was a man of many passions. Foremost among them was his desire to understand nature and to get at underlying truths. He was devoted to his research. "I wouldn't want to live if I didn't have my work," Einstein wrote to his friend Michele Besso.[1] He admitted that he "sold himself body and soul to science," but he didn't regret it.[2] Research was his life, and he devoted himself to it wholeheartedly. Still, Einstein had many other passions. Music was high on this list. He said that if he had not become a physicist he would have liked to have become a musician. "I often think in music. I live my daydreams in music. I see my life in terms of music. I get most joy in life out of music," Einstein once said to a reporter.[3] He particularly liked Mozart. "Mozart's music is so pure and beautiful I see it as a reflection of the immense beauty of the universe."[4]

Einstein's wish to unify physical science is also well known. He spent the last thirty years of his life in a desperate search for a unified theory of the universe, a theory that would explain all of physical science. In the end it eluded him, but his efforts were not in vain. Although few people took this work seriously when he was alive, the search for a "Theory of

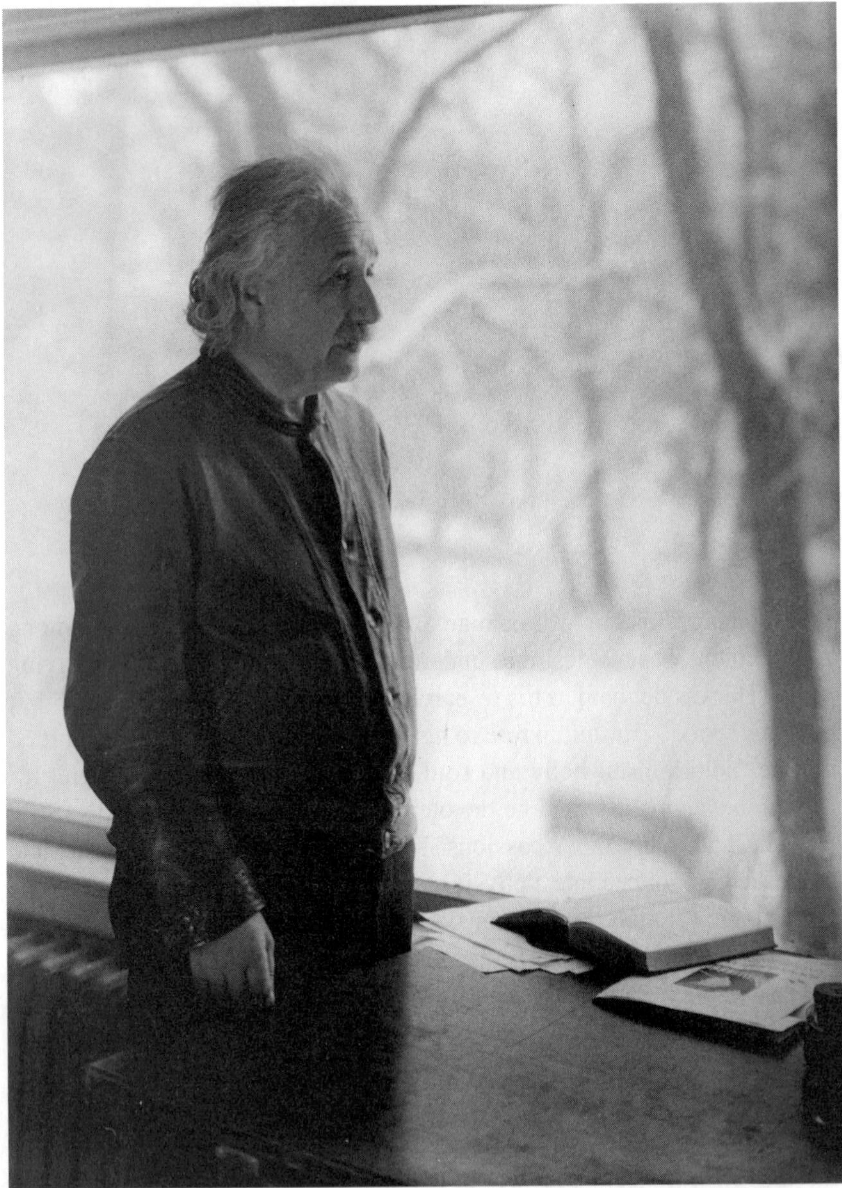

Fig. 1: Einstein at the Princeton Institute for Advanced Study

Everything" is now an important branch of physics, a branch that involves hundreds of scientists around the world.

In pursuing his passions, Einstein was sometimes seen as stubborn and unyielding. These traits were noted in both his quest for a unified theory of the universe and his rejection of quantum mechanics. In quantum mechanics the determinism of classical theory was replaced by probability and uncertainty. Einstein found this repugnant and refused to accept it. To him the universe had to be deterministic; in other words, a scientist should be able to determine its properties "exactly." As far as he was concerned, it was the only thing that made sense. One of his most famous quotes is, "God does not play dice with the universe."[5]

Einstein looked for simplicity in his theories, and he also had a desire for simplicity in his personal life. He had a disdain for money and never took advantage of the many opportunities for making money that were offered to him. He was generally uncomfortable and confused about the fame that was thrust on him, remaining unassuming and humble through it all. Indeed, his need for simplicity even extended to his dress; he preferred old, worn clothes to new ones, and he frequently went without socks.

Despite his fondness for simplicity, Einstein was not a simple man. He had a complex personality. Though many people think of him as a calm and avuncular person, Einstein was in reality a gentle man who felt things deeply and pursued his goals with great relish. He had an underlying intensity that he brought to everything he loved. Banesh Hoffmann refers to Einstein as a "creator and rebel"[6] in his biography of him, and indeed both of these traits, along with his strong feelings about the world around him, were evident from an early age. Einstein hated the militaristic teaching methods that were used in his Munich school; in fact, he hated all aspects of the military. This is no doubt why he became such a strong pacifist in later life.

Although he had a loving family, there was considerable turmoil in Einstein's early life. He was left in Munich to finish high school when his parents moved to Milan, which made him depressed. He didn't get along well with many of his teachers, and at university he was rebellious and headstrong, and certainly not a favorite among his professors. It was the first few years after his graduation, however, that were the lowest, most depressing years of his life. He couldn't find a job, and his parents were strongly against his marriage to his college sweetheart. For a time he felt worthless, but he was determined to succeed. Indeed, it was this determination, this driving force within him, that kept him going.

Although he was rebellious when young, as Einstein got older, his nature changed, at least on the surface. He had always been sure of himself, but when he was young, he felt that he had to prove himself, which made him appear cocky. After he finally did prove himself, he became calmer and more confident. But beneath the surface, his youthful passion and drive were always there.

Some of his personality traits can be traced to his mother's attitude toward him. She had strong feelings for him, but she did not smother him with love; she was more interested in his future and in encouraging him to be self-reliant. Her husband, Hermann, was easygoing and easily swayed by others, so she was determined that her son would not end up the same way. She was going to make sure that her son developed a strong backbone, and there's no doubt that she had an influence on his personality.

Einstein's father also had an influence on him, although it wasn't as great as his mother's. He was warmhearted and kind, and Einstein had great affection for him. Still, young Einstein saw his father's shortcomings and had no intention of following him into the family electrical business, even after he was encouraged to do so. Einstein felt sorry for his father after his business failures, and he knew that the pressure of his work was taking a toll on his health. He encouraged him to get out of the business, but it did little good.

Although Einstein enjoyed solitude, he had many close friends, and these friendships lasted throughout his life. Most of his closest friends were men, but he liked the company of women and enjoyed being around them. Strangely, though, for the most part he did not accept them as intellectual equals, at least not as far as science was concerned. It can be safely said that in addition to his passion for science, Einstein had a strong passion for women, even though they brought him considerable trouble at times. The four main women in his life were his mother; his sister, Maja;

and his two wives, Mileva and Elsa. (Later on his secretary played an important role in his life, too.) He also had a love interest when he was quite young, named Marie Winteler. His letters to Marie, Mileva, and Elsa convey strong feelings. To Mileva he wrote, "When you're not with me, I feel as though I'm not complete."[7] On another occasion, "Without the thought of you I would no longer want to live among the sorry herd of humans."[8] And to Elsa he wrote, "I must love someone. Otherwise it is a miserable existence. And this someone is you."[9]

Einstein's love for books and learning when he was young is also well known. He spent much of his time reading the popular science books of the day; he even tackled some very comprehensive and advanced ones, such as Immanuel Kant's *The Critique of Pure Reason*. Throughout much of his early life, Einstein preferred self-study to learning things in class, and even at university, he spent much of his time in self-study.

Einstein's peak of creativity came shortly after he graduated from university. With two students, Maurice Solovine and Conrad Habicht, he formed the Olympic Academy, and over the next few years, the three of them studied the great works of science. At this time, ideas were coming to Einstein in a flood, and he discussed them with his students. He was now dealing with some of the most important problems in physics. Discussing these problems with his students helped Einstein understand them better. One of the major ideas he was considering was related to light and electromagnetic waves. Several years earlier he had tried to visualize what it would be like to chase after a light beam and catch it. He saw that there were serious problems related to it, but for a period of several years, he continued to puzzle over the idea.

Then came the "miracle year" of 1905. During this year Einstein published five of the most important papers ever published in physics. Among them was his famous paper on special relativity that gave us a new picture of the role of space and time in the universe. It contained many strange predictions. But Einstein knew that special relativity was not complete; it applied only to straight-line, uniform motion, and he was soon attempting to extend it to all types of motion. It was at this time that he encountered what he later referred to as "the happiest thought of his life." This thought soon showed him that there was an "equivalence" between accelerated motion and gravity, which he used to formulate his famous "equivalence principle."

Within a short time he had completed his generalization of special relativity, now referred to as *general relativity*. The strain from the effort was so great, however, that he underwent a physical and mental collapse when the theory was finally completed, and he had to take to bed. A few years earlier he married Mileva, but the marriage soured and they separated. His cousin Elsa looked after him during this time, and a few years later the two were married.

General relativity's success pleased Einstein. We now know that in addition to explaining gravity, it predicted such exotic things as black holes, and it has also explained the beginning of the universe and its overall structure. Einstein eventually began to feel that it could be extended to cover the other major field of nature, namely the electromagnetic field. He was influenced by two attempts in this direction, one by Herman Weyl and the other by Theodor Kaluza. Both seemed reasonable and quite ingenious to Einstein, but both were eventually shown to be incorrect. Einstein soon got into the act. He was convinced that the two fields were just two manifestations of the same field, and over the years he made many attempts to bring them together. This was the beginning of his quest for a unified theory of the universe, which extended throughout the rest of this life. He had a firm belief in the unity of science. This is evident in his quote: "The grand aim of all science is to cover the greatest number of empirical facts by logical deductions from the smallest number of hypotheses or axioms."[10]

Einstein hoped his new theory would also explain the major particles of nature. As he struggled to find a theory, however, things got much more complicated. The basic fields of nature went from two to four, and the number of elementary particles mushroomed into the hundreds. Furthermore, a new theory—quantum mechanics—was put forward that appeared to explain things quite well. Strangely, although Einstein made important contributions to the theory early on, he had a hard time accepting it. He disliked the uncertainty associated with it. Nothing could be predicted with certainty; it could only be given a probability. "Quantum mechanics is worthy of regard. But an inner voice tells me that it is not yet the right track," he wrote.[11] Einstein worked hard to show that the new theory had flaws. One of his most famous papers (written with Boris Podolsky and Nathan Rosen in 1933) introduced what is now called the *EPR paradox*. In the end, though, quantum mechanics was shown to be valid. In 1965 John Bell of CERN (the European Organization for Nuclear Research) introduced an inequality that was designed to resolve the problem, and in 1983 Alain Aspect of the University of Paris used the inequality to show that the basic ideas of quantum mechanics were correct.

Einstein was in Germany during World War I and barely escaped the wrath of Hitler against the Jews just prior to (and during) World War II. Because of this, he developed a strong desire to bring about peace. He deplored war and did as much as he could in his later years to promote the idea of world peace. Although he had a very small role in the building of the atomic bomb, Einstein deplored its use. After World War II he campaigned for a world government, certain that it was the only way worldwide peace could be achieved. Although the United Nations was formed after the war, nothing ever came of Einstein's idea of a world government.

Einstein was not a pure pacifist, however. He knew that at times force was needed. "I am not only a pacifist, but a militant pacifist. I am willing to fight for peace. . . . Is it not better for a man to die for a cause in which he believes, such as peace, than to suffer for a cause in which he does not believe, such as war," he wrote.[12] He explained his pacifist views as follows: "My pacifism is an instinctive feeling, a feeling that possesses me because the murder of people is disgusting."[13]

Except for a short period in his youth, Judaism did not play a large role in Einstein's early life. With the advent of anti-Semitism just prior to World War II, however, Einstein began to acknowledge his religion publicly. "So long as I lived in Switzerland, I did not become aware of my Judaism. This changed as soon as I took up residence in Berlin. I saw how anti-Semitism prevented Jews from pursuing orderly studies, and how they struggled to secure a livelihood," he wrote.[14]

When the war was over, Einstein supported the creation of Israel as a refuge for Jews. He was such a strong supporter of the Hebrew University in Jerusalem that he was offered a position at the university, but he declined it. He was also offered the leadership of the new Jewish state, but he declined that, too.

Chapter 1

An Early Passion for Learning and Music

From an early age Einstein had a love of learning, but that was only part of his complex personality. He also had a passion for puzzles and mathematical problems, and his tenacity and stubbornness were evident even then: once he had started a problem, he refused to give up before it was solved. Furthermore, like Isaac Newton, who liked to build windmills, waterwheels, and sundials in his youth, Einstein loved to construct things. With his blocks and Anker set (something like the Tinker Toys many children play with today), he built many complex structures. But his favorite pastime was erecting many-storied houses with cards. Most of his friends had trouble building two- or three-story houses, but Einstein was able to build them to fourteen stories before they collapsed. He was delighted with these accomplishments and liked to show them off. In the final analysis, however, as a child, it was books that were the real love of Einstein's life.[1]

Einstein was born on March 14, 1879, in Ulm, Germany. But he didn't stay in Ulm long; in the summer of 1880 his family moved to Munich, which was about a hundred miles away and considerably larger in size. His mother, Pauline, was a tall, strong-minded woman with a

Fig. 2: Einstein with his sister Maja. He is about five years old.

broad nose, gray eyes, and a full figure. She loved music and managed to instill this love in her son. She liked to tease him when he was young. As he grew older, however, he became as strong-minded as she was, and it was inevitable that there would be clashes, and indeed there were. His father, Hermann, had a dark mustache that extended beyond his upper lip, a broad, firm jaw, and a slightly receding hairline. His hair was cropped short, and he wore a monocle, giving him the stern appearance of a Prussian army officer, but in reality he was kind and friendly, and was loved by almost everyone who knew him.

A year and a half after they arrived in Munich, Einstein's sister was born. She was named Marie, but everyone called her Maja. Einstein had a close relationship with her throughout his life, but when he was young, he was rather hard on her. On one occasion when he was about five, he threw a bowling ball at her, and on another he hit her over the head with a child's hoe.[2] When they first moved to Munich, they lived near the center of the city, but by 1885 they had moved to the suburb of Sendling. Their house was a large, two-story structure with a sun roof. Several large trees stood in the backyard along with many neglected flower beds.

Hermann Einstein had gone into an electrical business with his brother Jakob, and Jakob lived next door to them in an adjoining house. Jakob was a favorite of Einstein's; he was an electrical engineer, and Einstein could talk to him in a way that he couldn't talk to his father. His father didn't understand math and science, and Jakob did. Although Einstein felt close to his father, his interest in science and math drew him to his uncle.

Hermann and Jakob's business—a large, impressive building in which dynamos, armatures, arc lamps, and various types of electric meters were manufactured—was only a short distance away from their house. Jakob was the technical adviser and inventor; Hermann looked after the business and sales.

When Einstein was five, his parents hired a tutor, but it was soon evident that he wasn't an ideal student. He threw a chair at the tutor and chased her off during one of the sessions. Pauline didn't give up; she hired another and his schooling continued. At the age of seven he entered public school, where his tutoring paid off: he was able to enter at the

Fig. 3: Pauline Einstein (Albert's mother)

Fig. 4: Hermann Einstein (Albert's father)

second grade. He did not feel comfortable and generally kept to himself at elementary school. The school was Catholic, and he was Jewish, but his parents were not devout and never went to synagogue, so for them it was not a problem. Besides, the only Jewish school in Munich was a long distance away, and the Catholic school was within walking distance.

Einstein appeared to be slow when he first entered school. He didn't answer questions quickly in class, preferring to think things through before answering. Some of it was no doubt caution, since an incorrect answer would bring a wrapping on the knuckles. His parents, in fact, had worried because he began talking late. But it was not a serious problem. Rather than saying one or two words, he waited until he could talk in sentences. Indeed, he developed a habit of saying the sentence to himself before he said it out loud, and sometimes it was audible, so people thought he was repeating himself.[3] Despite this, he was always at or very near the top of the class. In August 1886 Pauline wrote, "Albert got his grades yesterday, he was again at the top of his class, he brought home a brilliant record."[4]

Einstein was the only Jew in his class, which, at about seventy students, was quite large by today's standards. Although he rarely joined in the games that other children played, there is no indication that he experienced religious discrimination, but he was teased and frequently referred to as a "bore." He was taught algebra by his Uncle Jakob about two years before he took it in school.[5] And Jakob had a unique approach to the subject, one that no doubt appealed to Einstein. He made a game of it, referring to x as a little animal that was unknown. "When we bag our animal, we pounce on it and give it its right name," he would say. His attitude toward Einstein was one of teasing, but he always encouraged him and was impressed with his talents. It's easy to imagine him saying, "I bet you can't solve this." And of course Einstein reveled in this; he loved a challenge and wouldn't give up until the problem was solved. He enjoyed the sessions with his uncle, and according to his sister, he would jump for joy when he had solved a problem. You can imagine him yelling, "See, I told you I could solve it." Writing about this period of his life, Maja said, "Persistence and tenacity were already part of his character and would become more prominent."[6]

On October 1, 1888, Einstein entered the Luitpold Gymnasium, which is roughly equivalent to our high school. He was nine and a half years old. A heavy emphasis was placed on Latin and Greek, with most of the class time being taken up by these subjects. In addition, however, he received instruction in German, French, mathematics, geography, literature, and science. Einstein didn't mind Latin because of its logical structure, but disliked Greek and most other languages, and eventually came to dislike the gymnasium in general. In particular he hated the militarism and strictness of the teachers and the rote learning methods. He later referred to his teachers in elementary school as "drill sergeants" and his teachers in the gymnasium as "lieutenants."

Fortunately, there was an outside influence who played an important role in Einstein's development. His name was Max Talmud. The Einsteins generally did not follow the usual traditions of the Jewish faith, but they did follow one tradition. They invited a poor Jewish student to their dinner table once a week, and the recipient of their generosity was Talmud.

It's interesting to note that while his family did not take Judaism seriously, Einstein did, at least for a while. According to German law he was required to take religious training, and since he was Jewish, it was Judaism for him. In elementary school a distant relative took care of it, but when he went to the Luitpold Gymnasium, his training continued at the school.

By the time Einstein was eleven, the religious teachings had begun to influence him. As he learned more and more about the Jewish religion, he became very disenchanted with his family and scolded them for not following its traditions. They did not observe the Sabbath, they did not pray at home, and they ate pork. For a year or two he took it upon himself to set an example; he would not eat pork, and he even went as far as composing and singing religions songs on his way to and from school. He had even planned on becoming bar mitzvahed, thereby becoming a full member of the Jewish community. This was to happen on his thirteenth birthday. The reason it didn't was that his religious feelings came to an abrupt end when he was about twelve. Science came into his life.

Talmud, who later changed his name to Talmey, visited the Einsteins each Thursday. He was a Jewish medical student at the University of Munich who was of Polish decent. At twenty-one, he was eleven years older than Einstein, but he was so impressed with Einstein's intelligence that he was soon treating him as an intellectual equal. For his part, Einstein was intrigued with being able to discuss math and science with a university student who was so much older than he. Einstein couldn't have wished for a better mentor; Talmud was soon spending hours with him, discussing math and science, and later philosophy.

It was about this time that Einstein got some of his class books early, just before the summer vacation. One of the books, a geometry book, attracted his attention, and soon intrigued him. He began working the exercises and looking for alternate proofs for many of the theorems in the book. He would show his solutions to Talmud each Thursday. Over the summer he worked his way completely through the book. It became his passion, and in his usual determined way, he solved every problem in it.[7]

It is perhaps strange that Einstein, who would not accept anything without proving it for himself, was not bothered by Euclidean geometry. At its foundation are five axioms that are accepted as self-evident. In short, they cannot be proved. But as Einstein later wrote, "[The] lucidity and certainty [of Euclidean geometry] made an indescribable impression upon me. That the axioms had to be accepted unproved did not bother me."[8]

Fig. 5: Max Talmud

Einstein later referred to the book as his "holy geometry book." There is still some uncertainty as to the identity of the book. Lewis Pyenson, in his book *The Young Einstein*, states that it was probably the geometry part of Sickenberger's mathematics texts, which was a separate book. Banesh Hoffmann, on the other hand, refers to it as Heis and Eschweiler's text on geometry in his book *Creator and Rebel*. When Talmud saw Einstein's interest in the geometry book, he brought him Theodor Spieker's book *Lehrbuch der ebenen Geometrie*, but it didn't have the impact of the first book. Over the next couple of years, Talmud continued to bring Einstein books, both on mathematics and physics, and later on philosophy.

By the time Einstein was thirteen, he was already beginning to study calculus. But he was becoming increasingly fascinated by physics and philosophy. Among the books Talmud brought him were *Force and Matter* by Ludwig Büchner and *The Cosmos: An Attempt at a Description of the Physical World* by Alexander von Humboldt.[9] They were the two best-selling popular science books of the day, so it's not surprising that he got them for Einstein. The books that may have had more influence, however, were those of Aaron Bernstein. They covered all areas of physical science and ran to twenty-one volumes. We do not know which of the volumes

Einstein read, but they were likely the ones on electricity, light, and the forces of nature. He no doubt also read the volumes related to astronomy.

Each of Bernstein's books was a series of essays.[10] In one, titled *The Wonders of Astronomy*, he described how Urbain Leverrier determined the position of the planet Neptune from its perturbation on Uranus. In another he described an imagined trip to the planets of the solar system, and of particular interest, he imagined the trip took place on a telegraph signal. Einstein later tried to imagine what it would be like to travel through space on a light wave, and Bernstein may have been the inspiration for these thoughts. Bernstein was particularly interested in telegraphic messages; earlier, he had invented a way of sending two telegraph signals over a single wire.

Bernstein also discussed the forces of nature. In an essay titled "The Secret Forces of Nature," he went into considerable detail on attractive and repulsive forces and their significance. Something that had to have profoundly excited and influenced Einstein was Bernstein's statement in these essays that many scientific problem had not yet been solved. He referred to them as challenges for the future. In regard to the magnetic field, Bernstein wrote, "Everybody feels that natural science is here just at the beginning of its scientific conquests and that there remains much, extraordinarily much to do."

Another of Bernstein's essays was on light. He discussed the speed of light and Isaac Newton's "Corpuscular theory of light." We know that Einstein referred to Newton's theory in his famous paper on the photoelectric effect, and this may have sparked some of his thoughts on the subject. Bernstein also discussed the force of gravity. Einstein, no doubt, read all of these essays with intense interest. Talmud later wrote that Einstein read the books with "breathless suspense."[11] Bernstein also discussed several experiments that could be performed at home. It is not known whether Einstein attempted any of them, but they may have been the source of his fascination with the laboratory when he later went to college.

Finally, Bernstein emphasized the importance of unity in science, and of course the unity of science eventually became one of Einstein's passions. The entire latter part of his life was a struggle to find unity in the forces of nature.

Einstein advanced so rapidly, particularly in mathematics, that

Talmud was soon unable to keep up with him. In an effort to hold his own, Talmud began directing the discussion to philosophy. He encouraged Einstein to read Kant's *The Critique of Pure Reason*, a book that is difficult for a university student to read, let alone a twelve-year-old boy. Einstein became fascinated with the book and continued reading and rereading it over the next few years. Talmud later wrote, "In all those years I never saw him reading any light literature."[12]

But Einstein's sudden exposure to the world of science had a profound effect on him. Upon reading science books, and Kant's book on philosophy, he began to see that there was a conflict between science and religion. Many of the biblical stories he had learned from his Judaism teachers could not be reconciled with science. One or the other obviously had to go, and for him it was religion. He didn't complete his bar mitzvah, and his religious phase came to an end.

It was about this time, however, that another passion came into his life—a lifelong love for music. Einstein's mother, Pauline, was an excellent pianist and was determined to instill an appreciation for music in both of her children. Einstein was given lessons on the violin, starting at about age five. Maja was given lessons on the piano. Einstein, who had a distaste for any type of rote learning, was not particularly attracted to music at first, yet he practiced his scales and études dutifully. Over the years he continued practicing, but his indifference continued. His mother kept encouraging him, because she hoped eventually to play duets with him. Finally, when he was about twelve or thirteen, Einstein discovered Mozart and fell in love with his sonatas. Their beauty and logical structure reminded him of mathematics. He couldn't get enough of Mozart and soon had learned several of his sonatas well enough that he could play them as duets with his mother.

Referring to music, he later wrote, "I believe, on the whole, that love is a better teacher than a sense of duty—with me, at least, it certainly was."[13]

In addition to Mozart, Einstein liked Schubert and Bach, but his mother's favorite was Beethoven. In later life he told his son Hans, "Music was one of the most important things in my life." His motto in regard to music was, "Listen, play, love, revere—and keep your mouth shut."[14]

Einstein also played the piano, according to Maja, even though he never took lessons. Most of what he learned he picked up on his own. He liked to sit at the piano and improvise, playing arpeggios with his left hand while he picked out simple melodies with his right. He also liked to explore different harmonies and chord transitions.

Throughout much of his life Einstein used music to relax his mind when he needed to think. This is illustrated in a story told by Charlie Chaplin in his autobiography. Elsa, Einstein's second wife, told Chaplin the story of the morning Einstein created his general theory of relativity.[15]

> The doctor [Einstein] came down in his dressing gown as usual for breakfast but he hardly touched a thing. I thought something was wrong, so I asked him what was troubling him. "Darling," he said, "I have a wonderful idea." After drinking his coffee, he went to the piano and started playing. Now and again he would stop playing and make a few notes.

He continued this for a half hour while he thought about the significance of his breakthrough. He went up to his room, and when he came down two weeks later, he had several sheets of paper in his hands. On them were the equations of his general theory of relativity.

When Einstein was fifteen, his life changed. Talmud had graduated, so he no longer had a mentor; nevertheless, he continued studying on his own. The big change, however, came about as a result of problems in the family business. Hermann had joined Jakob in an electrical business (at first it also included plumbing) when they first came to Munich in 1880, and for the first few years the business flourished. Indeed, by the early 1890s it was quite successful. Electricity was coming into its own, and it was an exciting time for the electric industry. Whole towns were being lit for the first time by electric lights, and for a while the Einstein brothers were right in the middle of the action. They supplied electric power to Schwabing, a suburb of Munich, and had a large display of electrical equipment at the Frankfurt Exhibition in 1891.[16] Over a million people attended the exhibition, with the Kaiser himself spending a day touring the grounds. They also supplied the power and electric lighting for two towns in northern Italy: Varese and Susa.

Of the brothers, Jakob was by far the more ambitious. Hermann was more content with what he had, but he was easily swept along by Jakob, who had a much stronger will. Jakob had invented a dynamo and wanted to manufacture it, along with other electrical equipment such as arc lamps and electric meters. They finally set up a large factory that at one point had almost two hundred employees. But they had considerable competition, mostly from large companies outside of Munich. Their downfall came when the city of Munich finally decided to convert to electric lights. As the only dynamo manufacturer in Munich, the Einstein brothers were sure that they had a good chance to get the contract. But there were several problems. First of all, their experience was mainly in direct current (DC), and when it was shown that long-distance transmission of electricity via alternating current (AC) was much less expensive than DC, they knew they had to make some quick changes if they were to compete with the larger out-of-town companies. And that required a lot of capital. They had already borrowed heavily from relatives, so they now had to mortgage their property and houses.

The brothers were going up against some of the largest electric companies in Germany: AEG of Berlin (the German Edison company), Siemens and company (Werner Siemens had earlier invented the first dynamo), and Schukert of Nuremberg. For several months there was serious wrangling among the companies, but in April 1893 the contract was awarded to Schukert of Nuremberg. Hermann and Jakob were devastated; they had a large factory, and it would have been kept busy if they had gotten the Munich contract. Their overhead was high, and there was little business left in Munich; in addition, they had mortgaged their homes. They had to do something.

After discussing the problem with their Italian representative, Signor Garrone, Jakob and Hermann decided to liquidate everything in Munich and move to Pavia, which was just outside Milan in northern Italy. Garrone convinced them that business was likely to be much better there. They had supplied electricity to two nearby towns, and there was the prospect of setting up a hydroelectric plant at Pavia. Garrone agreed to become a partner, and he put up some money. One of Pauline's relatives also agreed to loan them some, so they decided to move.

Einstein was surprised when the announcement came from his father that they were going to move, but he was pleased. He was glad because school was becoming an increasing burden for him, and he was beginning to hate it. A move would be a relief. But his parents had other plans. They told him that he had to stay in Munich by himself and finish gymnasium.

Einstein must have been shocked. He had three more years to complete gymnasium, which must have seemed like an eternity to him. And he no doubt argued with his parents about the decision, but their minds were made up. There were several reasons for their decision. He was unfamiliar with the Italian language, and they felt it was best to complete his schooling in Germany where he was familiar with the school and the language. Furthermore, he had a military obligation to fulfill when he graduated, and it was virtually impossible to get around it. Once he turned sixteen, even if he left the country, he had to return for the obligation, or he would be branded a deserter. He was only fifteen at the time, but would be sixteen in a matter of months.[17]

His parents made arrangements for him to board with an older woman. A distant relative, an aunt, would look in on him from time to time.

It's hard to judge his parents' reaction to leaving Einsten in Munich by himself for such a long time. He had strong ties to his mother, but she never smothered him with love, and her main concern was always his welfare. He also had strong feelings for his father, but Hermann generally went along with whatever his wife said, and she must have thought that it was best. There's little doubt, however, that Maja was devastated because she was very close to her brother.

To add to his misery, Einstein had to watch the destruction of his house and yard before he left. It was sold to a contractor, who immediately pulled out all the large trees in the backyard, then razed the house, and built several apartments in their place.

In the summer of 1893 Einstein waved good-bye to his family at the railroad station in Munich.

Chapter 2

Leaving Munich

Einstein was lonely and depressed after his parents left. He buried himself in his books, but he missed his family. They wrote to him, and although their letters were eagerly awaited, they usually left him melancholy and despondent. His mother wrote about their new apartment in Milan, their new life, and the beautiful countryside around the city. He also got letters from his cousins in Genoa telling him about the grandeur and beauty of Italy. He must have begun to think of it as a paradise. All in all, it was a joyless and depressing time for Einstein.

He immersed himself in the study of the philosopher, Immanuel Kant. Earlier he had read Kant's *History of Nature and Theory of the Heavens*, and it had intrigued him. He was fascinated by some of the ideas Kant put forward. One of them was the suggestion that our solar system formed from a gigantic nebular disk; another was that our galaxy, the Milky Way, was an "island universe" of stars, surrounded by millions of similar island universes. But the book he spent the most time on was *Critique of Pure Reason*. Near the beginning of this book, Kant gives a detailed philosophical discussion of space and time. "Time is not an empirical concept," writes Kant. In other words, it is not based on experiment. At another point he writes, "Time is nothing else than the form of the internal sense. . . . It is a subjective condition of our [human] condition." Statements like this must have made Einstein think. Time and space were to become central concepts in his theory of special relativity, which came about a decade later, and it's obvious that he was already thinking seriously about them. Space and time seemed to him, at first glance, to be simple, straightforward concepts. But was there some hidden meaning behind them? Were they more complex than they appeared to be? Einstein was sure that they were not fully understood.

Even though his studies occupied him, they didn't take the place of his family. "[He] was depressed and nervous," wrote Maja.[1] He had three years of school left, then he would be drafted into the army for another three years, and he hated everything about the military. Unlike most German boys at the time, Einstein had never dreamed of becoming a soldier, marching in parades, or fighting in a war. In fact, he dreaded the prospect. Once while watching soldiers in a parade when he was young, he said to his parents, "When I grow up I don't want to be one of those poor people."[2]

The school year began on September 10. Einstein would be going into his seventh year at the gymnasium, which is roughly equivalent to our grade ten. He had few friends at the school and disliked most of his teachers. It wasn't that he was a poor student; he was always near the top of his class. The grading system at the Luitpold Gymnasium was from 1 to 4, with 1 being the highest (roughly equivalent to our A). Einstein records were destroyed in World War II, but Dr. Wieleitner, a later principal, reported in 1929 that Einstein's grades were always high. He routinely got 1 or 2 in Latin and Greek, and even higher grades in mathematics. In his last year, however, he did receive a 3 in Greek, which is roughly equivalent to our C.

Einstein did have problems with some of his teachers. He particularly disliked one, but a few were friendly and encouraging. His homeroom teacher in his fourth year (1891/92) and sixth year (1893/94), and also his German literature teacher, was Dr. Ferdinand Ruess. Ruess did not require rote memorization, as many of Einstein's other teachers did, and he encouraged independent thinking. Under him, Einstein studied the well-known German authors Johann Goethe and Friedrich von Schiller. He thoroughly enjoyed Ruess's class, and Ruess instilled in Einstein a love for the German authors. Einstein's father also particularly liked Schiller and occasionally read aloud from his books to the family. Einstein remembered Ruess as a favorite teacher in later life. While passing through Munich one time, when he was successful and a professor at a large university, he decided to look him up and visit him. Ruess, however, didn't remember having Einstein as a pupil, and the visit was a bit of a disaster, with Einstein making a hasty exit.

Einstein also liked his mathematics teachers. For several years he had Joseph Zametzer, but in the last two years his math teacher was Joseph Ducrue. Ducrue was also his physics teacher. Einstein's major problem was his Greek teacher, Dr. Joseph Degenhart, who was also his homeroom teacher in his seventh and last year at the gymnasium. Degenhart made it clear that he had no use for Einstein.

Between September and December, Einstein became increasingly depressed, and his hatred for the school intensified. He felt abandoned by his parents. There is no indication that they invited him to spend Christmas with them, and he ended up spending it alone. It was the first time he had ever been away from his family at Christmas. His pent-up feelings of remorse and frustration were no doubt close to the boiling point when he visited his doctor sometime before Christmas for a minor ailment.[3] His doctor, Bernard Talmud, was an older brother of Max Talmud (it was Bernard who introduced Max to the Einstein family). It is not known whether Einstein had already hatched his plan for joining his family when he visited Dr. Talmud or whether it began with the visit. Anyway, he must have poured out his feelings to Talmud, and Talmud agreed to give him a letter saying that he was on the verge of a nervous breakdown and needed a rest, and it was best that he join his family in Italy. Talmud may have done this as a favor because the Einstein family had been good to Max, but he no doubt could see that Einstein was very depressed.

This was the first step in Einstein's plan, but he needed more. He wasn't sure how or where he would finish gymnasium, but he knew he wasn't going to finish it in Germany, and he wanted to go to university. He hoped one day to teach philosophy, and perhaps physics, and knew that without a college degree he couldn't. He therefore went to his mathematics teacher, Joseph Ducrue. At the time Einstein was studying algebra, trigonometry, and geometry in class, but he had already mastered differential and integral calculus by studying them on his own. He was well ahead of his class, and Ducrue no doubt realized this. It's reasonable to assume that Einstein may have asked Ducrue for help occasionally with his self-study of calculus. Einstein therefore asked him for a letter stating that he was beyond the mathematics required for a gymnasium diploma, and Ducrue gave him the letter. He was probably one of the few teachers who was disappointed to see Einstein leave.

THE SURPRISE

Einstein now had what he wanted and began preparing to leave. He would merely have to take the letters to his principal. To his surprise, however, something happened that hastened matters, or at least made them easier for him. As mentioned earlier, much of Einstein's misery centered on his homeroom teacher, who was also his Greek teacher. Not only did Einstein dislike him, but he resented having to take Greek, sure that he would have little use for it after he graduated. He knew the teacher disliked him, but he didn't care.

In the days before Einstein was to leave, he sat in the back of the classroom in his Greek class, paying little attention to what was being said. Every so often he would feel the teacher's eyes on him, boring into him with disdain and contempt, but it didn't worry him. He was leaving. When the teacher looked at Einstein, he would smile back.

He had little reason to smile, however, since earlier his father had asked this teacher what he thought Einstein should go into—in other words, what profession he was best suited for. The teacher replied, "It doesn't matter. He won't make a success of anything."[4]

To Einstein's dismay, the teacher asked him to stay after class. "You are a disturbing influence in the class," the teacher said to him when Einstein finally faced him.[5] Einstein objected, "I have done nothing." "You sit there smiling like a jackass. It destroys the respect of the rest of the class for me," the teacher replied. "It is best that you leave the school."

Einstein most assuredly had mixed feelings about the statement. He now had a good excuse for leaving, but he hated to think that he was being kicked out. Maja later commented on the incident by saying, "And in fact Albert Einstein never did attain a professorship of Greek grammar."[6] There is obviously a note of sarcasm in the comment.

Einstein's last trip was to the principal. He presented him with the two letters and told him that his Greek teacher had asked him to leave. His departure date is recorded as a few days after Christmas—December 29, 1894. He packed his belongings, said good-bye to his landlady, and was soon at the Munich railroad station. He had not told his parents he was coming. Furthermore, although he had hiked to many nearby towns, he had never taken a long railroad trip. Milan is roughly 250 miles from Munich, on the other side of the Alps, so it was a new experience for Einstein.

He certainly felt relief, sure that anything to come would be better that what he had been through over the past few months, and much better than what he would have to go through had he stayed in Munich. The Alps would have been covered with snow as he traveled through them and must have been a beautiful sight. But Einstein's mind was on other things. Despite his relief, he had to be worried. What would his parents think? What would they say to him? They would be disappointed, but he was going to assure them that he would go to university, and he hoped the letter from his math teacher would be helpful, demonstrating to them that he was already at the university level in mathematics.

One of the most prestigious universities in Europe was close to Milan—just across the boarder in Zurich, Switzerland. It was known as the Zurich Polytechnic or ETH (Eidgenössische Technische Hochschule). You didn't need a gymnasium certificate to attend it; all you needed was to pass the entrance exams. It is not known whether Einstein knew about the Zurich Polytechnic and its requirement when he made the trip to Milan, but he certainly learned about it shortly after he arrived, and he promised his parents he would prepare for the exams.

He knew that his parents would be confused when he presented himself at their door, and would try to get him to go back, but he was determined. Under no circumstances was he going back. And there was another problem: He had decided to give up his German citizenship. He wasn't sure how they would react to that.

ARRIVAL IN MILAN

It's easy to imagine the surprised look on the faces of his parents when Einstein appeared at their door in Milan. At the time, they were living in a large apartment in the center of Milan. "They were alarmed by his high-handed behavior," said Maja.[7] And his stubbornness must have annoyed them even more. "He most adamantly declared that he would not return to Munich," wrote Maja. There was considerable arguing for a while, but his parents soon resigned themselves to the fact that he was determined not to go back. There was, of course, still the problem of conscription, and Einstein was just as determined about it. His hatred of the military went back to his early youth, and he was determined not to spend three years in the German army. Yet, strangely, later when he became a Swiss citizen, he gladly reported for army duty. He told his father he was going to give up his German citizenship, but he had to do it before his sixteenth birthday. His father signed for him, and he was relieved of it formally on January 28, 1896. For the next five years he was stateless, but he did become a Swiss citizen shortly after his twenty-first birthday.

Soon after Einstein arrived, his father took him aside and asked about his future plans. He had already promised that he would study for the polytechnic exams and take them in the fall, but what his father was interested in was what he planned on studying. Einstein had become fascinated by philosophy after reading Kant, and he was also interested in theoretical physics. He told his father he would like to teach philosophy and, needless to say, it didn't go over well. It thoroughly annoyed his practical-minded father, who told him to forget this "philosophical nonsense."[8] He told him that he couldn't make a living teaching philosophy and should become an electrical engineer like his Uncle Jakob. This was a little ironic in that Jakob and Hermann's electrical business had already gone under in Munich and was facing problems in Milan. Einstein liked electricity and magnetism, but he wasn't interested in an applied profession. He enjoyed the abstract and the theoretical. He argued with his father but eventually gave in and said he would become an electrical engineer. Later, he said that he didn't study for the entrance exams as hard as he could have because of that commitment. He knew that if he passed them he would have to sign up for engineering and he didn't want to.

Einstein's life and personality changed dramatically after he ironed out the problems with his parents. A load was lifted from his shoulders, and he soon began enjoying life. "A freer life and independent work made a quiet, dreamy boy into a happy, outgoing, universally liked young man," wrote Maja.[9]

Einstein thoroughly enjoyed himself over the next few months. He made arrangements to help at his father's factory, but most of the time he spent studying and traveling. Within a few weeks he had made several friends, and with one of them, Otto Neustatter, he hiked to several of the neighboring towns and cities. He also visited art centers and museums in Milan and nearby cities. Einstein particularly enjoyed the Italian people, writing later, "I was so surprised when I crossed the Alps to Italy to see how the ordinary Italian, the ordinary man and woman, uses words and expressions of a high level of thought and cultural content, so different from the ordinary German. . . . The people of Northern Italy are the most civilized people I have ever met."[10]

He embarked on a rather long trip to Genoa to visit his Uncle Jakob Koch and his cousins. He took the train part way, then hiked almost sixty miles across the Alps on foot. The trip took him several days. Though Einstein disliked organized sports and refrained from participating in them, he didn't dislike physical activity. In fact, he was an avid hiker and mountain climber, and spent a lot of time hiking in the Italian and Swiss mountains. Later in the summer he vacationed with his family at the Alpine village of Airolo, which is directly north of Milan, in Switzerland. At Airolo, he met the cabinet minister, Luigi Luzzatti, and became friends with him. Luzzatti later became the prime minister of Italy.

Einstein continued studying for the polytechnic exams. Shortly after he arrived in Milan, he bought a copy of Jules Violle's comprehensive treatise on physics and spent much of his time studying it. By the end of summer he had gone completely through the book. According to Maja his extraordinary powers of concentration were already evident at this time. "In a large, quite noisy group, he could withdraw to the sofa, take pen and paper in hand, . . . and lose himself so completely in a problem that the conversation of many voices stimulated rather than disturbed him," she wrote.[11] In addition to physics, he also continued to study mathematics, but the exam was to cover more than just mathematics and physics; it would also cover botany, geography, languages, and history. Einstein did not like studying these subjects because they involved so much memorization, and he hated memorizing things. He therefore spent little time on them.

In addition to studying, he helped out in his father's business. In fact, on one occasion he was able to help Jakob solve a particularly difficult problem. Jakob had been working on the problem, which was related to electricity and magnetism (likely an electrical system of some type), for several days with an assistant. He asked Einstein if he could help, describing the problem to him. Over the next fifteen minutes, Jakob watched in amazement as Einstein quickly solved the problem. Later he said to another engineer, "Where I and my assistant engineer have racked our brains for days, this young fellow comes along and solves the whole business in a mere quarter-hour. He'll go far one day."[12]

MORE PROBLEMS

There was, however, a shadow looming over his newfound freedom and independence. Hermann and Jakob had built their electric factory at Pavia, about twenty miles outside Milan, and it wasn't a small factory. A watercolor painting of it shows a two-story main building along with many long narrow buildings.[13] It would have taken up an entire city block in any city, so their overhead had to be high. As difficulties began to compound in the late spring of 1895, the two families decided to move to

Pavia to be closer to the business. But problems continued when the Einstein brothers ran into difficulties with water rights for their proposed hydroelectric plant. The Pavia city council finally voted to sever all business relations with the Einstein brothers, so everything fell through, leaving them bankrupt again.

Not long after this, Hermann came to his son and told him the bad news: he could no longer support him.

THE ESSAY

Although Einstein spent most of his time preparing for the polytechnic exams, he refused to give up his personal interest in studying the ether and its relationship with the electromagnetic field. Their relationship was beginning to become an obsession with him. What was the link between the electric and magnetic fields, and how did it relate to the ether? The ether had been invented a few years earlier after James Clerk Maxwell had shown theoretically that electromagnetic waves should exist, and Heinrich Hertz had shown experimentally in 1888 that they do exist. The ether was the medium that propagated the waves. Einstein was sure that the relationship between the electromagnetic field and the ether could be determined by studying the potential states (energy states) of the ether in a magnetic field, and by measuring the resulting elastic deformations (changes in shape) and deforming forces.

During the summer of 1895, Einstein wrote an essay outlining a program for studying the problem. Toward the end of the essay, he wrote, "I believe that quantitative researches on the absolute magnitude of the density and the elastic force of the ether can only begin if qualitative results exist that are connected with established ideas."[14] In other words, he was convinced that careful experimentation was needed, in addition to theory. There was nothing really new in the essay; it centered around ideas that were prevalent at the time. Nevertheless, it was a considerable accomplishment for a sixteen-year-old.

Einstein sent the five-page essay to his Uncle Caesar (his mother's brother) in Stuttgart. Koch was a grain merchant who bought and sold grains on the international market. In the accompanying letter, Einstein wrote, "It is rather naive and imperfect, as might be expected from such a young fellow as myself. I shall not be the least offended if you do not read the stuff at all." Einstein also mentioned in the letter that he would be attending the Zurich Polytechnic in the fall, but he didn't mention that he had to pass the entrance exams to do so.[15]

It's unlikely that Caesar understood even the simpler parts of the essay, but he was impressed. He wrote back to Pauline praising Einstein.

While Einstein was studying for the exams, Pauline wrote to the Polytechnic Institute in Zurich for more details about it. She was dismayed when the answer came back: to take the exams you had to be eighteen years old. Einstein had just turned sixteen.

Chapter 3

First Love

Pauline, as you might expect, was not happy with the news that Einstein was too young for the exam, and being as strong-minded as she was, she wasn't going to sit by and do nothing. She immediately began looking into what could be done. The family had a friend named Gustav Maier in Zurich, and Pauline wrote to him, asking if he would be able to help. Maier had been a neighbor in Ulm. She asked him to check if there was a way around the age requirement for the exam. Maier, in turn, got in touch with the director of the polytechnic, Albin Herzog.

Pauline must have told Maier that Einstein was a child prodigy, which Maier passed on to Herzog. Herzog replied with the words, "It is not advisable to withdraw a student from the institution in which he has begun his studies even if he is a so-called 'child prodigy.'"[1] Herzog said, however, that he would overlook the age requirement if Einstein's talents and intellectual maturity were confirmed in writing. Einstein must have known that his letter from his mathematics teacher would help somewhere along the line, and indeed it did. The letter was forwarded to Herzog, and Einstein was allowed to take the exams.

THE EXAMS

The exams began on October 8, 1895, and extended over several days. Einstein journeyed to Zurich just prior to this, likely with his mother. They stayed at the home of Maier. It was Einstein's first visit to the city, and it must have impressed him. It was a picturesque city, framed by the Albin mountains in the distance. The older section of the city extended upward from the Limmat River, with suburbs reaching around the shores of Lake Zurich. Vineyards and orchards covered the slopes of the hills that led down to the lake. Overall, it was a beautiful setting.

The exam consisted of two parts: one on general information and one on scientific and mathematical knowledge. The general exam covered literary history, political history, natural sciences, and German. The scientific and mathematics exam included arithmetic, algebra, geometry, physics, chemistry, and technical drawing. The results of the exam were announced on October 14. Einstein and his mother were called in for a conference with Herzog, and Heinrich Weber, the professor of physics at the polytechnic, may have also been present. Einstein had done well on the mathematics and physics part of the exam. Indeed, he did well enough on the physics part that Weber invited him to audit his classes if he was to stay in Zurich. The general information part of the exam, however, was a different matter. Einstein failed it, and because of this, he failed the overall exam.

Einstein was disappointed, but not surprised. He knew he was weak in several areas and must have known when he completed the exams that he had done poorly in the general knowledge part. He later wrote, "[The exams] made me realize painfully the gappy character of my previous schooling."[2] Herzog recommended that he finish high school in one of the nearby towns. With the help of Maier, arrangements were made for Einstein to attend a school in Aarau, which was about twenty-five miles from Zurich. He was to stay with one of the teachers of the school.

TO AARAU

The trip to Aarau was one of mixed emotions for Einstein. The school year had already started, so he would be slightly behind. Furthermore, Aarau was in the German-speaking section of Switzerland, only a few miles from the German border, and Einstein worried that the school might be like the one he had left in Munich. As the train approached Aarau, however, his tensions must have eased. It was an enchanting village in a magnificent setting. The buildings were decorated with elaborate gables, wrought iron emblems, and delicately curved eaves, and every so often there were beautifully colored windows. Aarau consisted of a maze of narrow streets rising from the Aare River, and in the distance were the Jura mountains.

Einstein was met by Jost Winteler, and he would spend the next year in the Winteler home. Winteler had a kind and warm personality, and would no doubt have made Einstein feel immediately at home. The secondary school, which Einstein was to attend, had three grades: one, two, and three (which are equivalent to our grades ten, eleven, and twelve). Einstein had two options: he could enroll in the gymnasium, as he had in Munich, where the emphasis was on Latin and Greek, or he could register in the trade school. Both would allow him to go to the Zurich Polytechnic. As you might expect, Einstein opted for the trade school.

Upon registering at the school, Einstein had to take another series of exams to see where he should be placed. It's important to remember that he was still sixteen at this time, and at Munich he had only gone through what was equivalent to the first half of grade one at the Aarau school. As it turned out, the results of his exams allowed him to jump to the third grade, but he was classified as a "provisional" student with gaps in French and chemistry, which required him to take private lessons in these subjects. It's interesting that his cousin Robert Koch came to the school at the same time as Einstein, and although he was the same age, he was sent to the second grade. Koch stayed in a room next door to the Wintelers.

THE WINTELER FAMILY

Einstein was extremely lucky to be taken in by the Wintelers. They were a loving family who treated him as one of their own. The Wintelers had seven children, but their house was large enough that Einstein had his own bedroom. Jost Winteler was a teacher of history and philology at the Aarau secondary school, which was just across the street from their house. Intelligent and knowledgeable, he had studied at the University of Zurich and the University of Jena, and Einstein soon looked up to him with respect and admiration. Winteler's wife, Pauline, had the same effect on Einstein. Although she shared his mother's name, she was quite different. Warmhearted and kind, she treated Einstein like a son. He would sit and talk to her for hours about almost anything, and when he called people "Philistines," as he frequently did, she just laughed and took no offense. Even after he left, he continued to write long letters to her, beginning them with "Dear Mommy" and signing off with "Thousand greeting[s] and kisses." Indeed, soon after he arrived at the Wintelers, he was referring to Jost and Pauline as "Papa and Mamma."

Fig. 6: Jost Winteler

In addition to Jost and Pauline, the Winteler family consisted of four boys and three girls, and they all soon treated Einstein as a brother. According to Anna, the oldest girl, Einstein was "fond of conducting scientific conversations, yet he had a great sense of humor and at times could laugh heartily." She went on to say that he studied most evenings.[3]

One of the things that appealed to Einstein was that Jost and the entire family were so liberal-minded. They would frequently gather around the table after supper and discuss various topics, some of them quite controversial. At other times Jost would read to the family, just as Einstein's father had. Einstein and the Winteler children were encouraged to join in the discussions, and they could say whatever they felt, without fear of reprimand. This appealed to Einstein, and he enjoyed it. He had many things to get off his chest, and foremost was his deep-seated loathing of Germany and its militarism. To Einstein's delight, Jost shared his views. He had studied in Germany and over the years had developed a mistrust of German aims, although it wasn't likely as strong as Einstein's. Jost was seriously worried about German expansion. In later years, when Germany did expand under Hitler's aggression, Einstein remembered Jost's predictions and reminded several of his friends about them.

Why was Einstein so strongly anti-German? Part of it was the militarism he had experienced in the schools and the attention that was paid to the military and its importance through parades and so on. The rigidity of thought and learning that he had experienced may have also contributed. Considerable resentment no doubt also came because of his expulsion from the Luitpold Gymnasium. And further resentment may have arisen because of his father's business failure. The Einstein brothers' business was the only electrical business in Munich, but the city council went outside the city to award the contract for lighting the city.

Einstein had to have let considerable steam off in these discussions. This is something that was unthinkable in Germany, but with Jost as an ally, he must have felt some relief.

FIRST LOVE

The biggest change that occurred in Einstein's life, however, was an emotional one. He had always thrived on the attention of women, and he enjoyed flirting, but now for the first time he was in love. The object of his affections was Marie, one of the three Winteler girls. At eighteen, she was two years older than Einstein, but in many ways she was less mature than he was, and because of this, he treated her as if she were younger. She knew she was less intelligent than Einstein and was a little in awe of him.

This might seem a little odd since at this time Einstein was not a particularly strong student at school. He had been admitted conditionally, and with the exception of mathematics and physics, his early grades were not outstanding. But he was, indeed, mature for his age. We see this in a letter from Gustav Maier of Zurich to Jost Winteler. He wrote, "Albert Einstein is much more mature than his cousin [Robert Koch] and therefore less in need of supervision."[4] Koch and Einstein were both sixteen years old. Marie likely formed her opinion of Einstein from the discussions at the table and around the house. Einstein liked to talk about science, which was generally over Marie's head, and it must have impressed her.

Marie was lighthearted and fun to be with, and Einstein was delighted with the attention and love she showered on him. She was also a diversion from his long hours of study, and according to Robert Schulmann, who once saw a picture of her, she was the prettiest of the three Winteler girls. There are, unfortunately, no photographs of her in the public domain.

One of the main things that attracted them to one another was their mutual love of music. Marie played the piano, so they were soon playing duets together, probably some of the same ones Einstein had played with his mother. There appears to be no doubt that they were in love and devoted to one another. But in a household of ten, it was difficult for them to be alone. One of the few times this was possible was during the hikes that Jost conducted. He was an avid bird watcher and frequently took some of his students, along with members of his family, on hikes to search for rare birds. Einstein and Marie went along on many of these hikes, and occasionally they likely lingered behind the main group.

THE SCHOOL

The school that Einstein attended was across the street from the Wintelers' house. It was an old building that had once been used as a hospital. The halls and rooms were narrow and dark, and the laboratories were not well-equipped. But when Einstein arrived, construction of a new school next door had already started, and there was considerable excitement. It would open in the spring.

Einstein signed up for German, French, Italian, mathematics, physics, chemistry, natural history, geology, and drawing. To his dismay, he found that singing, group exercising, and compulsory military training were also on the curriculum, but Einstein found he could substitute violin lessons for singing, group exercising was optional, and military training was required only of Swiss citizens. He was also required to take extra lessons in French, natural history, and chemistry.[5]

Einstein knew he was deficient in some subjects and had to concentrate on making them up. But to his delight, he soon found that his worries about the militaristic style of teaching he had encountered in Munich were unfounded. The teachers were much more liberal and tolerant, and encouraged discourse with the students. Three of his main teachers were August Tuchschmidt, who taught physics (and was also principal of the school); Freidrich Mühlberg, who taught geology; and Heinrich Ganter, who taught mathematics. Along with Jost Winteler, these three teachers were strong role models for Einstein and no doubt influenced his decision to go into teaching. He compared their lives to those of his father and Uncle Jakob and decided he preferred the life of a teacher. He particularly liked the independence they had.

His grades over the first few months were: German, 2–3; French, 3–4; history, 1–2; mathematics, 1; physics, 1–2; natural history, 2–3; chemistry, 2–3; drawing, 2–3; and violin, 1 (the range is 1 to 6, with 1 being the highest).[6] Although none of his grades, with the exception of French, were considered poor, some of them were only average. Was there a problem? Interest certainly played some role, since Einstein only had a casual interest in subjects other than mathematics and physics. Nevertheless, he knew he had to do well in all subjects. Was Marie a distraction? It's unlikely that the time he spent with her had any effect on his grades. Probably the most important thing was that Einstein had a lot to overcome; he had jumped a grade and had to make up for it. Furthermore, his enthusiasm did not spread evenly over all subjects.

CONTINUING LOVE FOR MUSIC

In addition to studying hard, Einstein continued practicing his violin regularly, and his love for it, already high, increased even more. He would take his violin with him whenever he went visiting. Indeed, he took it with him so often that Marie started teasing him about it, referring to it as his "dear child." Einstein's playing also attracted his teacher's attention. In a music exam in the spring, the examiner wrote, "One student, by the name of Einstein, even sparkled by rendering an adagio from a Beethoven sonata with deep understanding."[7]

Some insight into both his music and his personality is provided by a classmate, Hans Byland. Byland wrote that he sat watching Einstein play one day and was amazed by the feeling he put into the music. To Byland he appeared to have a split personality: passionate and emotional when he played his violin, but prickly and cocky otherwise. "The sarcastic curl of his lip did not encourage philistines to fraternize with him," Byland wrote.[8] He went on to say that Einstein was cocksure of himself—a person who voiced his opinions, regardless of whom they offended. There's no doubt that he could be impudent, but many teenagers suffer from this, and certainly the Wintelers enjoyed his company.

How important was Einstein's music to his intellectual development? That's an interesting question. Psychologists tell us that music has a definite effect on intellectual development, and with Einstein's deep love for music, we would be surprised if it didn't have an effect. The earlier anecdote of him playing the piano to relax while he thought about his new theory indicates a creative connection within him between music and his research. Many of the great scientists of the past had a deep love for music. Werner Heisenberg, for example, who gave us the first formulation of quantum mechanics, was an excellent pianist and loved music. Edward Teller, the father of the hydrogen bomb, was an excellent violinist. Others such as Niels Bohr also enjoyed music.

This brings us to the question of how good a violinist he was. According to several people, Einstein was an excellent sight reader who could play relatively complicated pieces by sight. His repertoire was quite extensive, including Mozart, Beethoven, Schubert, Schumann, and Bach. But the consensus is that he was far from a concert violinist; he was merely a good amateur.

CHRISTMAS AT THE WINTELERS AND BEYOND

As Christmas approached, Einstein showed how much he enjoyed the Winteler family. It was expected he would return to Milan and spend Christmas with his own family, but he chose to stay with the Wintelers. By now they had become his second family, and he was close to all of them. He knew that Christmas in Pavia with his own family would have been dull compared to Christmas with the large and exuberant Winteler family. And indeed, with so many children, and Marie nearby, the Christmas vacation was likely a joyous time for him. Furthermore, he had now developed a bond with the Swiss people and the small village of Aarau. He would recall both with fond memories in later life. It's safe to say that the year he spent at Aarau was one of the happiest of his life.

By now Einstein had written several letters to his parents about how happy he was, and this pleased them. Both Hermann and Pauline wrote letters to the Wintelers near Christmas, expressing their gratitude for the care he was getting. "It's a great relief to know that my son is under such loving care which is not only concerned with his physical well-being but also promotes his intellectual and inner life. . . . At this stage the heart is most receptive to a good model and I am convinced that your good influence will leave a lasting effect," wrote Hermann.[9] It is clear that Hermann had a tremendous respect for Jost Winteler, and was in awe of his intelligence and knowledge.

Despite his feelings of well-being, Einstein's grades had not improved much by Christmas. Jost Winteler was worried about them and felt that he could do better; he sent Einstein's grades to his family. Hermann replied that he was used to seeing a few "not-so-good grades along with very good ones" in Einstein's reports, and he wasn't worried.[10] He was sure that Einstein would improve.

The first time Einstein would visit his family would be during spring break, which came in April. By then his mother and Maja knew all about his new love, and he was teased as soon as he got home. Pauline, as usual, was eager to show off her son's violin playing to the ladies of Pavia, and he was soon playing for them. But Einstein quickly tired of this. Writing to Marie, he said that he preferred playing duets with her, complaining that "they [the ladies] expect both perfection and breakneck speed." The letter he wrote to Marie is replete with strong emotion, as were his other letters to her. "Many, many thanks sweetheart for your charming little letter, which made me endlessly happy," he wrote.[11] "I was now made to realize to the fullest extent, my little angel, the meaning of homesickness and pining. But love brings such happiness." Pauline also sent her regards at the end of the letter, assuring Marie that she had not read the letter she had written to Einstein. Pauline grew quite fond of Marie, and certainly hoped that Einstein would one day marry her. The feelings of the Winteler family were similar. Pauline and Marie actually exchanged letters a little later on.

Einstein enjoyed the time with his family, but was anxious to get back to Aarau. While he was gone, the new school was completed and was now ready for dedication and occupation. It was a significant improvement over the old building: a three-story structure of brick and stone with a large arched entryway and a clock tower on the top. The new physical laboratory was one of the best of its kind in the country. It contained alternating-current motors, dynamos, batteries, several modern galvanometers, and other meters. The adjoining shop was equipped with a milling machine, a lathe, and an endless variety of tools. Einstein would spend a lot of time there.

The dedication took place on April 26, 1896. It was a beautiful spring day, and the ceremonies were held outdoors. The main speaker was Dr. Tuschmidt, Einstein's physics teacher and principal of the school. He gave a long speech on the value of education and the higher principles of life.

They had barely moved into the new school when the May Day celebrations were upon them. Einstein was asked to participate in a mock sword fight. He had already developed his pacifist tendencies and abhorred any type of fighting; nevertheless, he reluctantly agreed. On the day of the fight he arrived at the field in a black fencing outfit. On his hat were a skull and crossbones. It was no doubt the only time he ever had a weapon in his hands. As it turned out, just as the fight was about to begin, it started to pour rain and it had to be called off.

Einstein did much better in the spring exams in late May. With the exception of French and technical drawing, his grades improved significantly. He did well enough in geology, in fact, to be selected as one of twenty students asked to accompany Professor Mühlberg on a field trip to Mt. Säntis, in the Appenzel region of the eastern part of the country. Despite Mühlberg's curt, abrupt manner, Einstein liked him; furthermore, he admired him as a scholar. Mühlberg had published extensively over the years and was an active researcher.

Mt. Säntis is 8,141 feet high. When Mühlberg and his group arrived in June, it was dull and rainy. Einstein wasn't wearing hiking boots, and as the rain continued, the trail, which was narrow and steep, grew quite treacherous. Einstein slipped and began sliding toward a steep drop-off, and likely would have been killed had he gone over it. Luckily one of his classmates stretched out a walking cane, and Einstein was able to grab on to it. It was a close call, and it's safe to say that if he had gone over the edge, the history of physics would have been different. Fortunately for science he didn't.

CHANGING RELATIONS WITH MARIE

There's no doubt that Einstein was smitten with Marie, and for almost a year he believed he was in love. But gradually he began to lose interest in her. There's some indication, in fact, that he was losing interest even before he left Aarau. When he went to the polytechnic in Zurich, she left for a temporary teaching job in Oldenberg, a small village to the northwest of Aarau. They continued to exchange letters, but Einstein's heart wasn't in it; his letters became fewer and farther between. Amazingly, however, he still sent his dirty laundry to her to wash and send back to him. After not hearing from him for a long time, she began to worry and wrote to his mother asking if anything was wrong. Pauline assumed it was just laziness on his part and told Marie as much.[12] But Einstein had decided to break off the affair. Unfortunately for Marie, he didn't have the nerve to tell her directly and just stopped writing to her. He did, however, write to her mother apologizing for the grief he had caused. Pauline Winteler didn't blame him; she understood and felt it was probably best for Marie to end it now. In later years, Marie wrote, "We loved each other sincerely, but it was an entirely ideal love."[13]

Despite the breakup, Einstein continued corresponding with and visiting the Winteler family. His sister, Maja, also attended the Aarau school and stayed with the Wintelers from 1889 to 1892, and eventually married one of the Winteler boys, so Einstein ended up being related to them. In addition, one of his best friends at the Zurich Polytechnic married Anna, the oldest Winteler girl.

Why did Einstein decide to break up? It's hard to say. One of the reasons may have been the approaching final exams. He was determined to do well in them and didn't want any distraction. But it seems unlikely that this was a serious problem. It's much more likely that he just grew tired of her. Furthermore, as we will see, by the time he finally broke off relations, his interest was directed elsewhere.

CONTINUING PASSION FOR PHYSICS

Even though Einstein studied hard for the final exams, his interest in the problems of physics that were beyond the curriculum never diminished. It was during this time that he began to use "thought experiments," a technique that he would employ frequently. In one of these thought experiments, he pictured a light beam in a moving media such as water. How did the motion of the transporting medium affect the speed of light? Would you add the speed of the medium to the speed of light to get its overall speed? Unknown to Einstein, this experiment had already been performed in 1853 by Armand Fizeau, who used water as the moving medium. Fizeau, in fact, had obtained a result that puzzled him.

In another thought experiment Einstein wondered what it would be like to travel alongside a light beam. What would happen, for example, if you traveled with the speed of light, and observed the light wave. Einstein realized that it would no longer be a wave; in short, it wouldn't be time-dependent, and this appeared to be a serious problem to him. As we now know, this was a first step toward his special theory of relativity.

Ideas such as this from a sixteen-year-old may seem a little frivolous, but they weren't. As it turned out, he was actually considering serious physics problems that most of the well-known scientists of the day didn't even realize were problems.

THE *MATURA* EXAMS

While Einstein was studying for his final exams, referred to as the *Matura*, his father's business in Pavia was being liquidated. Hermann had decided to return to Milan and set up another electrical business. Einstein tried to talk him out of it, but to no avail. This time Jakob didn't become a partner; he went his own way and took a job with another electric company. Einstein was distressed by the business failure and the effect it had on his father, but he tried not to let it affect his studying.

The exam consisted of a written and an oral part, with the written part being given on September 18, 19, and 21. Several essays were included on the written part: a synopsis of one of Goethe's plays, an essay on the galvanometer, an essay on glaciation, and an essay in French titled, "My Future Plans."[14] Despite the fact he got his lowest grade for the French essay (it was riddled with spelling errors and grammatical errors), it is one of the most interesting parts of the exam from our point of view. In the essay he said he planned on enrolling at the polytechnic in Zurich to study mathematics and physics, with the aim of becoming a teacher. He said that he preferred the theoretical parts of the sciences. "They are, most of all, my individual inclination for abstract and mathematical thinking," he wrote.

Einstein did exceptionally well in the exams, with 6s in mathematics and physics and 4s and 5s in most other subjects (the system had been changed so that 6 was now the highest grade). The only subject he did poorly in was French, with a grade of 3. The French teacher, in fact, protested his passing. The oral exams were held on September 30. It was customary for two representatives from the Zurich Polytechnic to be in attendance. Interestingly, one of the two was Albin Herzog, the director of the polytechnic, who had advised Einstein to finish high school at Aarau after he had failed the entrance exam.

Overall Einstein got a grade of $5\frac{1}{3}$ out of 6, which was the highest of the ten students taking the exam. He had obviously come a long way during the year: from being a provisional student with deficiencies to the highest in his class.[15]

THE FUTURE

Einstein was now ready to go to the Zurich Polytechnic. The year at Aarau had been an extremely successful one for him; he had matured considerably and had enjoyed being in love for the first time. But, as we saw, he broke up with Marie. As we will see later, the breakup had an effect on him, but it had a much more serious effect on Marie. She was devastated, and it took her a long time to recover. She eventually married and had two sons, but for the most part she led a rather tragic life. She was divorced a few years after she married. In later life, she tried to get Einstein's help in coming to America, but she never came. One of the greatest tragedies of her life was to occur only a few years after Einstein left. One of the Winteler sons, Jules, came home from a trip to America. He was a cook on a merchant ship. Unknown to the family, he had become deranged. He shot and killed Pauline Winteler and the husband of one of his sisters, then killed himself. The incident was a shock to Einstein, since he had always been especially close to Pauline (he sent a letter of consolation to the family), but it must have been an even greater shock to Marie, who was very vulnerable. She lost both her mother and a brother. Toward the end of her life, she spent some time in a sanitarium and died in one in 1957.

Chapter 4

Student Days and a New Love

Having passed his *Matura*, Einstein was ready to go to the polytechnic, but there was a problem. With Hermann's financial woes, there was no money to pay for his tuition. Pauline decide to write her sister-in-law, Julie Koch. The Kochs were wealthy, and Pauline hoped they might help. Einstein had never cared for his Aunt Julie; to him she was pompous and vain, but he would gladly accept her money if it was offered. To Pauline's delight, Julie said she would send Einstein one hundred Swiss francs a month. It wasn't a large amount, but it would allow him to live moderately.

Einstein arrived in Zurich to begin his studies at the polytechnic in the middle of October 1896. The first thing he had to do was register. About a thousand students were registering, so the halls were crowded. Almost all the students were male, and most were registering for engineering. Einstein was nearly a year younger than most of them; he was still six months short of his eighteenth birthday. Eighteen was the legal age for entrance to the school, but he had been exempted.

Einstein registered in department VIA, the school for teachers of mathematics and physics. Although he was majoring in physics, he would also be able to teach mathematics and astronomy. With the help of some of the professors, he worked out a study program. He would be taking only mathematics the first semester; his first class in physics—a mechanics course—wouldn't come until the second semester. He would start off with differential and integral calculus, analytic geometry, descriptive geometry, and philosophy. He had already studied calculus on his own and was quite proficient in it, so he wasn't worried about the calculus class.

Within the first week or so, he found lodging at Unionstrasse 4 in the boarding house of Frau Henrietta Häge. It was within easy walking distance of the polytechnic.

NEW FRIENDS

Besides Einstein, there were four other students in his class; three were mathematics majors, and he and another student were physics majors. One of the math majors was Marcel Grossmann. He had dark wavy hair, a small neatly trimmed mustache, a slender build, and had gone to school in Basel. His father manufactured agricultural equipment. Einstein soon became close friends with Grossmann. Over the four years they attended the polytechnic, they spent a lot of time sipping coffee together at the local cafes, discussing their studies and other things of mutual interest. Grossmann was one of the first to see Einstein's potential, and he developed a great respect for his talents. He took Einstein to his home at Thalwil on Lake Zurich to meet his parents, and is reported to have said to his father, "Einstein will be a great man some day."[1] When Einstein began skipping classes, Grossmann came to his rescue by lending him his notes. Einstein later referred to them as a "lifesaver." Grossmann would become a lifelong friend and in later years would give Einstein valuable help in formulating his theories.

Einstein also became close to one of the other mathematics students, Jakob Ehrat, who, like him, was Jewish. They would spend hours together discussing the problems that Jews encountered. Einstein admitted, however, that there was little discrimination at the polytechnic, and their dis-

Fig. 7: Marcel Grossmann at the polytechnic

cussions usually centered on Jewish problems in other parts of the world. The other mathematics student was Louis Kollros from La chaux-de-Fonds in the Jura, near the French border. The last of the students, the other physics major, was female, and this must have been a surprise to Einstein, because it was quite unusual for him to have a girl in his classes. All his classmates at Aarau had been male.

Einstein's love of music brought him more friends. Soon after arriving in Zurich, he heard that there was a musical get-together at the home of Selina Caprotti each Saturday afternoon. Musicians and people interested in listening to music gathered to enjoy a musical afternoon. Einstein began attending these sessions soon after he heard about them. At one of the first that he attended, he met Michele Besso, a mechanical engineer six years older than he, who had graduated from the polytechnic a few years earlier and was now working at a small nearby town. He had been born near Zurich, but his family moved to Trieste, Italy, when he was young, and he had attended school there. After an argument with one of the teachers, he was expelled and had to finish high school in Rome. He soon became a close friend of Einstein's, but the relationship was strange. Einstein had little respect for Besso's easygoing, aimless lifestyle and lack of ambition, often referring to him as a "schlemiel." Nevertheless he developed great affection for him, and soon realized that he had a sharp mind. Furthermore, to Einstein's delight, Besso was intensely interested in physics, and eventually he became an excellent "sounding board" for many of the ideas Einstein was exploring.

Sometime after Einstein met Besso, he introduced him to Anna Winteler, the oldest of the three Winteler girls, and Besso married her within the year.

MILEVA

Einstein's interest was soon piqued by the one woman in his class, the other physics major, Mileva Marić.[2] She was from Vojvodina, on the Danube River, in the north of what was Yugoslavia. When she lived there, it was known as Hungary. Her family now lived in Kac, near Novi Sad. Mileva, the oldest of three children, had a younger sister, Zorka, and a brother. Her father, Milos, had been in the army, but was now a civil servant, and was fairly well-off. Mileva was his favorite, and he referred to her fondly as "Mitza." She was born with a deformity in her hip that caused her to limp. Strangely, her younger sister, Zorka, suffered from the same problem. Mileva was quite self-conscious of her limp, and it made her shy and withdrawn. As a Serbian, she was also dark complexioned, which was looked down upon in parts of Europe, including Zurich. She attended the high school in Novi Sad where she did particularly well in mathematics and physics. It's interesting that Milos had to get special permission for her to sit in on the physics class, but despite this and being the only girl in the class, she got the highest grades.

About this time, she became ill. Because of this, and because the opportunities for girls in Serbia were limited, she moved to Zurich. The mountain air was better for her health, and she was admitted to the girls' high school in Zurich in 1894. In the spring of 1896, she passed the *Matura* and was ready to go to university. Although she was at the top of her class in mathematics and physics, she was also interested in medicine, particularly psychology, so she went to the University of Zurich, where she signed up for medicine.

She quickly tired of medicine, however, and the following semester she decided to transfer to the Zurich Polytechnic, where she signed up to become a teacher of mathematics and physics. She was twenty-one—three and a half years older than Einstein—and older than anyone else in the class. Einstein took classes alongside Mileva for the first two semesters and got to know her quite well, but there was little sign of romantic interest, particularly during the first semester. At the end of the second semester, however, he did go on a hike with her.

Fig. 8: Michele Besso, about the time he got married

As in the case of Marie, music helped foster their relationship. Mileva played the piano and sang. She had begun taking piano lessons when she was eight, so she was a fairly good pianist, and according to Einstein and others, she had a lovely singing voice. He was soon playing duets with her. During the first semester, Einstein was still writing to Marie and had not formally broken off with her, but he had lost interest. He must have compared Marie to Mileva, and indeed they were quite different. Mileva was serious, whereas Marie was lighthearted and easygoing. Mileva was determined and knew what she wanted, and had struggled for what she had achieved. Marie was much less ambitious and not nearly as intelligent as Mileva. On the other hand, it seems that Marie was much prettier than Mileva, but it didn't help her. Einstein had made up his mind.

Most people found Mileva sullen and moody, someone who said little but seemed to notice everything. She had deep-set eyes and long dark hair. Her friends described her as "plain," with one girlfriend actually using the word "ugly." Her photos don't indicate this, and Einstein certainly had no complaint about her looks. Some of his friends, however, were surprised that he was attracted to her.

Although to the outside world Mileva appeared gloomy, she presented

Fig. 9: Mileva Marić
at the polytechnic

a different side to her girlfriends. All of them, like her, were from the Balkans. In their presence she laughed and enjoyed herself—something she seldom did in public. Einstein also saw this side of her at first; she was teasing and coquettish in their early relationship, but later this changed. From the beginning, we know from his letters that she had a temper, and Einstein respected it.

To her credit, she was very orderly and brought some order into Einstein's life. In addition to her ability in mathematics and science, she was a good cook and seamstress, and sewed many of her own clothes. Einstein, on the other hand, was not only disorganized and disordered at times, but also absent-minded and forgetful. She helped him in this respect.

A MATHEMATICAL ATTITUDE

In class Einstein was also not the ideal student; furthermore, he soon developed a strange attitude toward mathematics.[3] Calculus presented few problems for him, but he soon found that there was a lot more to mathematics than calculus. It was split up into many distinct and different branches. He was soon taking classes in analytic geometry, geometry of numbers, descriptive geometry, projective geometry, theory of functions, potential theory, calculus of variations, differential equations, and several other branches of math. To some degree these classes overwhelmed him, and he began to develop a distaste for mathematics, particularly the more abstract branches.

Einstein admitted later that he had excellent math teachers, and he didn't take advantage of them. They included Hermann Minkowski, Adolf Hurwitz, Carl Geiser, and Wilhelm Fiedler. Minkowski was a well-known researcher, who had published extensively by the time he came to the polytechnic. His greatest contribution, and the one that made him world-famous, was to come later when he developed a four-dimensional formulation of Einstein's theory. Unfortunately, he was a poor teacher, which turned Einstein off. He liked to emphasize the abstract and never used examples in class, much to Einstein's disappointment. Most students found his lectures dull, and they certainly didn't appeal to Einstein. Einstein took many classes from him, but skipped many of his lectures. He even took classes in applied physics from him, but according to Einstein, Minkowski never used any examples from physics and never explained the physics concepts (e.g., force, energy, momentum) that he used.

Einstein finally came to the conclusion that most of the mathematics that was taught by Minkowski and others was irrelevant to physics. All he really needed, he felt, was basic calculus, simple differential equations, and algebra. The rest was superfluous, and he treated it as such. He studied the abstract math, but his heart wasn't really in it. Years later when he was searching for a higher dimensional geometry to help him in his formulation of general relativity, Grossmann reminded him that they had taken a course in it from Geiser. Einstein barely remembered the class.

Einstein's feeling at the time are summed up in a paragraph he wrote many years later. "I saw that math was split into many specialized areas, each of which would take up the short lifetime that is granted to us. I thus found myself in the position of Buridan's ass, which could not make up its mind between one bundle of hay and another."[4]

With such an attitude, it is perhaps surprising that Einstein took so many math classes. He actually took as many as Grossmann, Ehrat, and Kollros, and they were math majors. Aside from a few by Minkowski and Geiser, however, he skipped most of the classes. Minkowski once said to him, "Einstein, you are a lazy dog."[5]

Einstein eventually came to regret his attitude toward math. As he became more deeply involved in research, he began to realize that he needed a lot of

Fig. 10: Einstein as a student at the polytechnic

the abstract mathematics he had once rejected. "I could have got a good mathematics education, but I didn't," he wrote later in life.[6]

"HERR WEBER"

Einstein was disappointed that he would be taking only math classes during the first semester. His first physics class didn't come until the second semester, and to him it was hardly physics. It was a large class in mechanics—140 students—most of whom were engineers, which was taught by Albin Herzog, and it was directed toward engineering. It included strength of materials, elasticity, and dynamics. Einstein referred to it in a letter that he wrote at this time. "Herzog spoke very clearly and well on strength of materials, and somewhat superficially on dynamics, but that's to be expected in a 'mass' class."

Mileva left the polytechnic at the end of the second semester. She told Einstein sometime late in the second semester that she wouldn't be returning the next semester. The announcement no doubt surprised him. We're not sure what his feelings for her were at the time, but there is considerable evidence that he was interested; furthermore, it must have amazed him when she said she was going to the University of Heidelberg. Women could only audit classes at Heidelberg, and she would get no credit for any of the classes she took. Despite any advice Einstein may have given her, she withdrew from the polytechnic on October 5, 1897, and went to Heidelberg.

It's hard to say what her reason was, and there are indications that she was still uncertain what she wanted. She had spent one semester at the University of Zurich in the medical school, two semesters at the polytechnic, and now she was going to Heidelberg. There's no evidence that her leaving had a serious effect on Einstein, even though she gave no indication that she would ever be coming back. Several months later, however, when she wrote to him, telling him she was thinking of coming back, he seemed excited and strongly encouraged her to do so.

In her absence, Einstein began his first "real" physics class. It was

Fig. 11: Heinrich Weber

taught by Weber, and Einstein enjoyed it. Weber was an excellent teacher. "Weber lectured on heat . . . and with great mastery. I look forward from one of his lectures to the next," he wrote to Mileva.[7]

Weber had been at the polytechnic about twenty years before Einstein appeared.[8] He was forty-six years old and had obtained his doctorate at the University of Jena. After graduation he worked as Hermann Helmholtz's assistant at the University of Berlin for three years. Helmholtz became his mentor, and Weber corresponded with him for years.

Weber's research interests were optics, heat, and electricity, and he published several papers in these areas. Later his interest turned mainly to electrotechnology. In 1888, however, his research ceased; he now had a new interest that took up most of his spare time over the next four years. The polytechnic was building a new Physical Institute, and Weber would direct the building and equipping of the labs. So, while important advances were being made in physics—Heinrich Hertz's confirmation of Maxwell's electromagnetic waves, in particular—Weber ignored them and knew little about them.

Weber spared no expense in equipping the labs, and they were among the best in Europe. One of his first duties after the institute was finished was to hire an assistant. Upon Helmholtz's recommendation, he hired Jean Pernet, a decision he would regret. Pernet was dissatisfied almost from the beginning. He wrote to Helmholtz complaining that his teaching load was far too high (compared to Weber's), and he didn't have nearly as much lab space as Weber. Pernet was thoroughly frustrated with his position, yet strangely he stayed on, and before long Weber and Pernet thoroughly disliked one another and hardly spoke. This was the situation when Einstein came to the polytechnic. He had both men as lab instructors.

MILEVA'S RETURN

Einstein and Mileva wrote to one another several times while Mileva was in Heidelberg. Einstein wrote her a four-page letter shortly after the beginning of the semester, and she replied to it in late October. He had told her not to reply until she got bored. In her letter she said she "had waited and waited for boredom to set in, but until today my waiting has been in vain."[9] She was shut in by thick fog that day. At this stage there appeared to be little in the way of romance between them. Referring to a student Einstein had told her had dropped out of school because of a love interest, she wrote, "It serves him right; what's the point of falling in love nowadays anyway."

By February, Einstein had learned Mileva was going to return to the polytechnic, and he appeared to be quite excited. "I'm glad that you intend to return here to continue your studies," he wrote.[10] "Come back soon; I'm sure you won't regret your decision. I'm convinced that you will be able to catch up rather quickly." He went on to tell her about the courses she had missed: Hurwitz's course on differential equations and calculus of variations, the second semester of Herzog's course on mechanics, Weber's course in physics, and Fiedler's course on projective geometry.

Einstein spent the spring break with his family in Milan. When he returned in April, he met Mileva, and over the next few weeks, he helped her organize a study program for making up the material she had missed. Einstein was quite enthusiastic about helping her, and they soon began studying together. About this time he came across a book by Paul Drude of the University of Giessen on the ether and electromagnetic fields, and he was soon spending much of his time studying it.

In late April he became ill and was unable to leave his room for several days. He spent most of his time in bed reading Drude. When he recovered, he continued studying with and helping Mileva, but late in the semester he realized he would soon have to start studying for the intermediate exams. They were given halfway through the program and would come the following October, just before the beginning of the winter semester. Everyone was required to take them, and Einstein knew he would have to do some serious studying. He had skipped many lectures and was behind in his classwork.

With all the material she had missed, Mileva knew she wouldn't be ready for the exams and decided to delay them by a year. At the end of the semester, Mileva returned to her home near Novi Sad, but Einstein stayed on in Zurich with Grossmann to study for the exams. Einstein had already developed a distaste for cramming, but he had no choice, so he studied hard. After a short trip home in September, he returned in early October for the exams. They would be entirely oral and would last for several days.

Einstein was surprised when the results were announced.[11] Of the four people taking the exams, he got the highest grade: 5.7 out of 6. Grossmann was second with 5.6. Einstein was no doubt slightly embarrassed; he had skipped so many lectures, yet Grossmann had let him use his notes, and he had outscored him. Without the notes, Einstein knew he wouldn't have done nearly as well.

With the exams over, Einstein was relieved; he could now get back to his reading. He still hadn't finished Drude's book. Just before classes began, he moved to a new boardinghouse on Klosbachstrasse. His landlady was Frau Markwalder. The landlady's daughter, Susanne, an elementary school teacher, lived with her, and Einstein soon got to know her. Despite his relationship with Mileva, he was soon playing duets with Susanne, whom he liked to flirt with and tease.

PASSION FOR THE LAB

Einstein was looking forward to his third year because he would finally be taking labs. Two labs were on his curriculum: Weber's electrotechnical lab and one called "Physical Exercises for Beginners" that was taught by Jean Pernet. Within weeks Einstein was enjoying Weber's lab and spending considerable time in it, frequently neglecting lectures as he had done previously. The lab was a delight to him.

It might seem strange that Einstein, whose major interest was theoretical physics, would be so absorbed in laboratory work. But Weber's labs were exceedingly well equipped, and Einstein enjoyed the contact with experiment. He realized that a thorough understanding of experimental measurement and technique was of great value to a theoretician. "I worked in Professor Weber's physical laboratory with fervor and passion," Einstein wrote later.[12] He had continuous contact with Weber in the lab because Weber supervised all his labs personally and was very conscientious about them. Einstein actually took several lab courses from Weber over the next two years, and his grades in them were uniformly excellent—mostly 6s.

Einstein's second lab turned out to be much less delightful. Although he had looked forward to it, he was soon disenchanted. The main problem was the short, stocky instructor, Jean Pernet. Einstein disliked him almost from the beginning, and the feeling was mutual. Part of the problem was Pernet's personal frustration at the time, but he also felt that Einstein didn't give him the respect he deserved. Because of these difficulties, Einstein began skipping the lab, which only made things worse.

One of the major things that bothered Pernet was that Einstein never followed the written instruction he handed out. Indeed, on several occasions Einstein crinkled up the instruction sheet and threw it in the wastebasket. It's easy to imagine Pernet's anger after seeing his student doing this. He complained about it to an assistant one day. "He always does something different from what I have ordered," said Pernet, angrily. "Yes, he does indeed, Herr Professor, but his solutions are always right and the methods he uses are always of great interest," replied the assistant.[13]

Pernet's annoyance finally got the best of him, and he reported Einstein to the director of the institute for "neglect of duty." Einstein wasn't happy, and when he faced Pernet, Pernet said, "Einstein, you are insolent and arrogant. Furthermore, you're hopeless at physics. For your own good you should switch to medicine or law. Physics is too difficult for you."[14] After a brief silence, he continued, "Why don't you switch?"

"Because, Herr Professor, I have even less talent in medicine and law. Why shouldn't I try my luck in physics," replied Einstein.

Einstein got his revenge for the remark a little later. He saw a female student arguing with Pernet. After Pernet left, Einstein walked up to the student and said, "He's crazy, isn't he." The student turned out to room at the same boardinghouse as Mileva; her name was Margarete von Uexküll. Einstein offered to write up her lab for her. She was hesitant at first, but finally agreed to let him do it. The next time she saw Einstein, she told him that after Pernet graded it, he handed it back to her saying, "See I told you that you could do a good job if you set your mind to it." They both had a good laugh. Uexküll reported that Einstein also helped other students in the same way.

But in the end it was Pernet who won. He gave Einstein a 1 for the lab, the only failing grade he got while at the polytechnic. This is strange in that he got almost all 6s from Weber in his labs. The major problem was the clash of two strong and quite different personalities. Pernet's emotional state at the time left much to be desired, and anyone as arrogant as Einstein surely must have struck him the wrong way. In addition, Einstein spent most of his time in Weber's lab, and Pernet undoubtedly looked upon him as Weber's student, and with his dislike of Weber, he wasn't going to do Einstein any favors.

OTHER PASSIONS

Einstein loved working in Weber's lab, but aside from it, he had several other interests while at the polytechnic. It was here that he had his first experience with sailing, and it would become a lifelong passion. Einstein was, in general, not a physical person and generally disliked exercise, but he did like hiking and mountain climbing, as long as the mountain wasn't too steep. And with Lake Zurich at his doorstep, he soon began sailing in the afternoons. According to Susanne Markwalder, who sometimes accompanied him, he would carry a small notebook, and when the wind died down, he would take it out and start doing some calculations. He later used sailing much as he used music, in other words, to relax him during long, intense struggles with difficult problems, so he could focus and think more clearly. Occasionally, in fact, he would come upon the solution to a problem that he had struggled with for days or even weeks, after spending some time sailing.

Einstein also began smoking a pipe when he was at the polytechnic. He enjoyed sitting in a cafe smoking his pipe, with a pad of mathematical equations before him. Indeed, during most of the time he was at the polytechnic, he had a standing appointment with Grossmann to drink coffee and discuss physics and other things on their minds. He enjoyed talking with Grossmann. They had a different view of physics and mathematics, however. Einstein preferred physics to mathematics and looked upon mathematics as a tool, and nothing more. He was, as we saw, a little apprehensive about it. Grossmann, on the other hand, loved mathematics and wasn't sure why he had to take physics. He was sure that it was never going to be really helpful to him. On one occasion, however, he assured Einstein that he had discovered that physics was useful after all. He said, "At one time I seriously worried when I used the toilet after somebody. I thought I might catch something from the seat if I sat on it while it was still warm. Physics has shown me there is no danger. The warmth is just the heat from their ass."[15] Einstein must have doubled over in laughter at this.

Einstein loved food, although it can't be said that he ate well while in college. He didn't drink alcohol, but loved coffee and drank it by the gallon. He frequently referred to Mileva as his "fellow coffee guzzler." One of his favorite snacks was pastries, but he also liked sausages. In general, however, his college diet left much to be desired; sometimes when he was involved in a problem, he would forget to eat, and at other times a meal would consist of a quick trip to the pastry shop. This caused him stomach problems later in life.

DISAPPOINTMENTS

Although Einstein had grievances with Pernet and some of his other teachers, his major disappointment came from Weber. The most exciting discoveries in the world of physics at the time centered around James Clerk Maxwell's theory of electromagnetism. A few years earlier Hertz had made some monumental discoveries regarding the electromagnetic waves that Maxwell had predicted, and the announcement had taken the world of physics by storm. Weber, however, had been busy building his institute at the time and had not kept up with the latest discoveries, and because of this he knew little about the details of Maxwell's theory. He therefore completely ignored everything that Maxwell, Hertz, and others had discovered, much to Einstein's dismay. To Weber's credit, however, he did encourage students to do outside reading—beyond the class.

This was a major setback for Einstein, but it wasn't the only problem. In addition, Weber said very little about the foundations of physics, and he gave little or no instruction in the techniques of theoretical physics—Einstein's first love. These areas were not within his realm of expertise; he was primarily interested in electrotechnology and measurement techniques.

So, while Einstein started out with enthusiasm for Weber's lectures, he eventually became disenchanted and developed a resentment toward Weber. In turn, Weber became disappointed with Einstein; he had expected great things from him, but when he found out that he was independent and arrogant, he was not happy. To add to his annoyance, Einstein referred to him as "Herr Weber" instead of the much more respectful "Herr Professor." One day Weber said to him, "Einstein, you're brilliant, but you have a serious problem. Nobody can tell you anything. You don't listen to anybody."[16]

When Einstein finally realized that he wasn't going to get any guidance from Weber in what he really loved, namely, theoretical physics, he turned to Minkowski. Several of the courses he took from Minkowski in the last year were strongly related to theoretical physics (e.g., mechanics of rigid bodies, theory of spinning tops). But Minkowski was also of little help. He ignored any application of his mathematical methods to physics. Near the end of Einstein's last semester, Minkowski gave a lecture on capillarity (the rise of water or other fluids in a narrow or very fine tube). When Einstein walked out after the lecture, he turned to his classmate Kollros and said, "That is the first lecture on mathematical physics that we have had at the Poly."[17]

In his disappointment, Einstein turned to self-study, and over his last year or so at the polytechnic, he spent a tremendous amount of his time in self-study. This is, indeed, where he learned his theoretical physics. He studied the works of Hermann Helmhotz, Ernst Mach, Rudolf Kirchhoff, Heinrich Hertz, James Clerk Maxwell, and Paul Drude, and was fascinated with what they had to say. Besso introduced him to Mach's book *The Science of Mechanics*, which contained some radical ideas, two of which were that space and motion might not be absolute (as predicted by Newton), and that Newton's laws may not be correct under all conditions and should be reexamined. Helmhotz also appealed to Einstein in suggesting that everything not verifiable by experiment or observation be rejected. And finally, Kirchhoff was also skeptical of Newton's laws; he was not sure they could explain some of the recent discoveries. Einstein was pleased with this skepticism and reveled in it.

To add to Einstein's woes during his last year, he was seriously injured in the lab. His right hand, the one he used for bowing his violin, needed several stitches after he hurt it in the lab. It has been reported that an explosion in Pernet's lab caused the injury, and it came after Einstein threw Pernet's instructions for the lab in the waste basket. This would, indeed, make a good story, but he was taking both Pernet's and Weber's labs at the time, and although he referred to the injury in two different letters, he didn't say which lab it occurred in.

In the spring break of his third year (1899) Einstein returned to his home in Milan. This time he carried a photograph of Mileva with him. He showed it to his mother and became nervous when she studied it for a very long time. He finally said, "She's very clever." Little did he know that this long, critical look was an omen of things to come.

Chapter 5

The Women in His Life

Although Mileva was the central woman in Einstein's life, she was far from the only one. Einstein loved to be around women, loved to talk to them, loved to flirt with them, and rarely missed an opportunity to do so. As you might expect, therefore, he became close friends with several women, even though he was engaged to Mileva. One was Susanne Markwalder, the daughter of his landlady during his third year at the polytechnic. He played duets with her and even went sailing with her. We don't know what her feelings for him were, but she was quite curious to see Mileva, and when she appeared one evening to study with Einstein in the living room of the boarding house, Susanne made sure she got a good look. She told her mother she wasn't impressed; she found Mileva quite plain and certainly not much of a match for Einstein. Of course, there may have been a slight touch of jealousy in her analysis.

An amusing incident occurred one night while he was playing duets with Susanne. One of the boarders of the house asked if she could bring her two daughters to listen to Einstein and Susanne as they practiced. Einstein always enjoyed playing for an audience and was pleased to be asked. That evening the three women sat on the sofa while Einstein and

Susanne began playing. Soon all three of them were knitting, and the clicking and banging of their needles began to become quite noticeable as the playing continued. Then every so often one of them would drop a stitch and sigh loudly as the others laughed softly. Einstein soon became agitated as it became obvious to him that they weren't paying much attention to the music, so without a word he stopped playing and put his violin in its case.[1] "Why are you stopping?" asked the mother of the two girls. "Because we would not dream of interrupting your work," replied Einstein as he picked up his case and walked up to his room.

Susan may have been embarrassed, but we don't know her reaction. Anyway, later in the day Einstein apologized to her, and that evening he played outside the window of the women's room.

A PASSION FOR KNOWLEDGE

Einstein's habit of skipping classes didn't stop during his third year, but he was now heavily involved in a self-study program. While at Aarau, he had spent a great deal of time thinking about the ether: the mysterious substance that was believed to propagate light. He had thought up an experiment that would allow him to test it, or at least test how fast we were moving through it, assuming it existed. During his third year he began thinking about the experiment again, and eventually approached Weber, asking permission to perform it. He wanted to measure the velocity of the earth relative to the ether using a semisilvered mirror and two thermocouples (devices for measuring temperature). The semisilvered mirror would allow some of the light from a light source to get through and some to be reflected. If the mirror was set up so that some of the light from a light source passed through parallel to the motion of the earth in its orbit, and some was reflected in an antiparallel direction, Einstein reasoned that there would be an energy difference between the beams. He assumed he could measure this difference using the thermocouples.

Unknown to Einstein, several experiments of this type had already been performed, the most famous of which was the Michelson-Morley experiment.[2] It's interesting to compare Einstein's proposed experiment with the Michelson-Morley experiment, which was conducted at the Case Institute of Technology in Cleveland in 1887. Albert Michelson and Edward Morley also used semisilvered mirrors, and they had a light source that traveled in the direction of the earth's orbital motion, but they also had a beam that moved perpendicular to it. Their experiment differed from Einstein's in that it was based on an effect called *interference*, which arises when two beams pass through one another. The waves of the two light beams interfere with one another, and Michelson and Morley were able to use an apparatus that Michelson had just invented, called an *interferometer*, to measure the change in interference. Their experimental apparatus was mounted on bedrock in an effort to avoid any external vibrations or other sources of error. Einstein's experiment was considerably less sophisticated, and it is unlikely that even if he had performed it, he would have seen any difference in the energy associated with the two beams using his thermocouples.

At any rate, Weber turned him down. He may have felt there was little chance of the experiment succeeding, or he may not have realized its importance. Einstein, as you might expect, was upset, but soon afterward he read about a conference on the ether that had taken place at Düsseldorf, Germany, in September 1898. Wilhelm Wien had presented an overview paper on the ether that was published in *Annalen der Physik*, the most prestigious physics journal in Europe. In it Wien looked into the problem of whether the earth carries the ether along with it (as it carries its atmosphere) or whether it passes through the ether. In particular, he discussed thirteen different experiments in which an attempt had been made to detect the motion of the earth relative to the ether. The Michelson-Morley experiment was among them.

Einstein wrote to Wien about his experiment, but it is not known if he received a reply. Nevertheless, his interest in the ether and the electromagnetic field continued through his third year at the polytechnic. During this time he worked closely with Mileva, and their relationship continued to blossom. He even began referring to her boarding place as "our place," and instructed his mother to send packages of "goodies" directly to it. He studied several books through the year. A new book, with new insights, was always a joy to Einstein; he looked forward to each one with a feeling of excitement, and his need for a deeper understanding was being met mostly by studying these books. He knew lectures were important, and that he was going to have to pass the finals (which were little more than a year away) if he was to get his diploma, but they were far less enjoyable than delving into books and getting the answers on his own. Besides, he always had Grossmann's notes to rely on. He particularly disliked rote learning and memorization; to Einstein, understanding was the most important thing.

The summer semester was over in late July, and Einstein was to join his mother and sister at a resort at Mettmenstetten, which was just south of Zurich. Mileva was returning to her family's farm at Kac to study for the intermediate exams, which would be administered in October.

OFF TO PARADISE

Einstein boarded a train for the short trip to Mettmenstetten. His mother and Maja were staying at the Hotel Pension Paradise. He would spend from August 1 to September 11 at the resort. He had several books with him to keep him busy over the six weeks, and he had soon settled into a routine of studying his books in the morning, and hiking and playing his violin in the afternoon. He hiked with the hotel owner, Robert Markwaller, and also with Maja. Indeed, he climbed Mt. Säntis, which was not far away, with Maja. This was the same mountain he had almost lost his life on earlier.

Despite the love Einstein felt for his mother and sister, he eventually became disenchanted with them. It seemed that the laughter of earlier days was gone; they seemed so serious, so preoccupied with insignificant things. They were so different from Mileva; he could talk to her about physics, about his hopes and research in a way he couldn't with them. They were becoming boring. Little did he know that most of the tension was caused by Mileva. Pauline had not seriously worried about his relationship with Mileva at first; she was sure it wouldn't last. But here he was beginning his last year, and he was still going with her. Pauline was now beginning to worry, and it was affecting their relationship.

In Einstein's first letter to Mileva, he wrote, "Here in Paradise I live a nice, quiet philistine life with my mother hen and sister."[3] He told her about the books he was reading and the ideas he was developing. He knew she would be "pouting and fretting," as he put it, if she knew what a good time he was having hiking, loafing, and studying what he wanted. She was "cramming," and he was sure she was hating it, so he ended his letter with, "You, poor girl, must be stuffing your head with gray matter, but I know that with your divine compassion, you'll accomplish everything with a level head."[4] In an effort to reassure her, he told her he wished he could be with her.

Despite the peaceful atmosphere at the Hotel Paradise, distractions abounded, and Einstein eventually began to resent them. His mother had a large number of friends at the hotel, who made frequent visits. Their chattering got on his nerves and made it difficult for him to study; furthermore, his mother frequently asked him to play his violin for them.

There was a pleasant diversion, however, and it was the sister-in-law of the owner, Anna Schmid. She was seventeen, and Einstein liked flirting with her. She may have been one of the few of his female friends with whom he didn't play duets, but there was considerable music-making in the evenings, and she joined in. Just before he left, she brought her autograph book to Einstein and asked him to write a verse in it. He wrote the following:[5]

Little girl, small and fine
What should I inscribe for you here?
I could think of many things
Including a kiss
On your tiny little mouth.
If you're angry about it
Do not start to cry.
The best punishment is—
To give me one too.
This little greeting
Is in remembrance of your rascally friend
Albert Einstein

Despite the many diversions, Einstein did get considerable work done. He was reading Hertz's book on the ether and electrodynamics. Hertz had presented a formulation of electrodynamics, but there was a problem with it. To implement it, he had to make an assumption about the motion of the ether—whether it was stationary relative to the earth or moving along with it. Hertz liked the latter version, in other words, the assumption that the ether was dragged along with the earth. He no doubt realized, however, that this would eventually lead to problems. Einstein expressed considerable skepticism of this view. Writing to Mileva, he said, "I'm convinced more and more that the electrodynamics of moving bodies as it is presented today doesn't correspond to reality, and that it will be possible to present it in a simpler way."[6] Although he admired Hertz, and what he had done, Einstein felt that he was on the wrong track, and that "beam experiments" would eventually be needed to decide the fate of the ether. With such speculation it's fair to say that he was well on his way to his special theory of relativity, or at least he was taking the first steps.

JULIA NIGGLI

While Einstein was at Mettmenstetten, he received a letter from Julia Niggli of Aarau. He had met her in Aarau and had played duets with her. Niggli was about six years older than Einstein, and therefore about twenty-six at this time. They had discussed the possibility of getting together to play duets, but they had not made a definite commitment. To his surprise, however, this letter was mostly about her love life. An older man had been courting her, and there had been some discussion of marriage. She wanted Einstein's advice, and he was eager to give it, but it may not have been what she expected. He told her not to expect happiness with someone else, even if she was in love with him, because he knew what the man would be like.[7] After all, he was a man, and he knew their motives and feelings. Her lover would be high spirited one day, sullen and cold the next, and it's likely he would also be unfaithful, ungrateful, and selfish. Einstein claimed he knew all this from experience, which was, of course, quite an admission. And it makes one wonder if he was thinking of Marie when he wrote it. Incidentally, Niggli never did marry this older man, and in fact never did marry throughout her life.

In the letter, Niggli also invited him to come to Aarau to play duets with her. But he couldn't tell his mother he was going to Aarau to visit Julia; he needed a better excuse. As it turned out, the Wintelers had a visiting scientist staying with them, a Professor Haab. Einstein therefore told his mother he was going to visit Professor Haab. He had an additional worry, however: there was a chance he might bump into Marie. She was still teaching at Oldberg, which was only a short distance from Aarau, and there was always the possibility she might come home on short notice.

Einstein went to Aarau, and luckily for him, he didn't bump into Marie, but he did have a hard time getting Julia Niggli to play duets. She was more interested in asking his advice about her love life. While he was there, he asked Julia to visit him at the Hotel Paradise. She was shocked by the request. Einstein was amused and laughed, "We'll be chaperoned," he said. "My mother and sister are there."

To his surprise, while he was at Aarau, he got a letter from Mileva that had been forwarded from Mettmenstetten. She was trapped on her family's farm at Kac; an outbreak of scarlet fever and diphtheria in the nearby town of Novi Sad forced her and her family to stay on the farm. She reported her "cramming" was proceeding slowly. "Fiedler is my biggest headache and his material is the hardest to master," she wrote.[8] Then she asked him for advice about the exam. "Do Fiedler and Herzog ask specific things, examples, or do they ask only general questions?"

Her letters were starting to get gloomy, and it bothered Einstein. He preferred perky, uplifting letters, like the ones she had written from Heidelberg. "Don't you feel sorry for me?" she wrote. It was obvious to Einstein that she was more worried about the intermediate exam than he had been. Furthermore, he and Grossmann had studied together for the exam, and they had spent a lot of time joking with one another. Mileva was studying alone. He tried to comfort her in the letter he wrote back, but it did little to alleviate her anxiety.

THE FINAL YEAR

On September 11, Einstein, his mother, and Maja returned to Milan. Einstein continued his self-study, and he continued trying to reassure Mileva through his letters. The intermediate exams were now only a couple of weeks away for her. "You shouldn't let this little exam bother you so much. It should be easy for you—especially with such harmless competition," he wrote.[9]

Maja had decided to go to the teachers' college in Aarau, and she would be staying at the Wintelers, as Einstein had. She wanted him to accompany her, and he did. He had heard, however, that Marie was now at home, so he didn't linger when he dropped Maja off. He wrote to Mileva telling her that he was going to Aarau, but he also told her something that wasn't likely to make her happy. He wrote, "The critical daughter with whom I was so madly in love with four years ago is coming back home. For the most part I now feel quite secure in my high fortress of calm. But I know that if I saw her a few more times I would certainly go mad. Of that I am certain and fear it like fire."[10]

Fortunately for him he did not encounter Marie, and he was back in Zurich on the same day. By now Mileva had completed her intermediate exams, and he was anxious to find out how she had done. She had not done well, but she had passed. Of the six people taking the exams, she ranked fifth, much lower than Einstein, who had ranked first when he took the exam. She got 5.05 out of 6. In physics, however, she got 5.5, the same as Einstein; her downfall was mathematics.

Shortly after Einstein got settled, he did something he had been looking forward to for several years: he applied for Swiss citizenship. For several years he had been saving twenty francs each month toward citizenship, and in October he sent in the necessary application forms. In late November the local police submitted a report of good conduct, but it would be more than a year before all the paperwork was complete and he actually became a Swiss citizen.

In early November 1899, Einstein moved to Unionstrasse so he would be closer to Mileva, who was still on Platterstrasse. His new landlady, surprisingly, was Frau Häge, the same landlady he had had when he first came to Zurich. Einstein and Mileva studied together as much as possible during the year, and it was during it that they decided to get married. But both Einstein and Mileva knew there was a problem: what would his mother think? At Mettmenstetten, Einstein thought he heard his mother talking about Mileva in a negative way and had reported it to Mileva. He tried to soften the blow by saying, "Maybe I only imagined it." Nevertheless, it was a serious worry for both of them.

Einstein became even more worried when he returned home at Christmas. His mother said nothing about Mileva; she never asked about her and never mentioned her name. To Einstein this was a little unnerving.

OTHER WOMEN IN HIS LIFE

Einstein had close friendships with Susanne Markwalder, Julia Niggli, and Anna Schmid. Each of these friendships was based largely on music. But he also had many other women in his life. Mileva frequently had tea in the afternoons with several of her girlfriends. Einstein would sometimes attend and would frequently offer to walk one of them home, and on at least one occasion, he carried Marie Rohrer's books from the library. Another of Mileva's girlfriends, Milana Bota, was quite impressed with Einstein at first and had plans of playing duets with him, but Mileva quickly put a stop to it, and for a while Mileva and Milana were not on speaking terms.

Eventually Milana and another girlfriend, Ruzica Drazic, began to resent Einstein because Mileva was spending so much time with him and ignoring them. Much to Mileva's surprise, they announced that they had decided to move out. Einstein got even by writing a sarcastic poem and sending it to them.

It was about this time that Einstein and Mileva began calling one another "Dollie" and "Johnnie." Einstein, in fact, seemed to love pet names, and in his letters he referred to Mileva by many different ones including, "little witch," "Sweet little one," "Dear little one," "Dear Kitten," "Dear Child," and "naughty little sweetheart."[11]

It wasn't that Einstein didn't have male friends. He was very close to Besso, Grossmann, and also to Ehrat. In addition, he was friendly with Friedrich Adler. Adler studied chemistry and physics, but his major interest was politics. His father was the head of the Socialist Party in Austria, and he was also an ardent socialist. In 1916 Friedrich Adler assassinated Austria's prime minister in the belief that it might shorten the war. He was sentenced to death, but Einstein, who by then was quite famous, helped get his life spared.

WORRIES BECOME PANIC FOR MILEVA

Both Einstein and Mileva were worried about Pauline's reaction to their impending marriage. Late in the winter semester, Mileva found out that one of her girlfriends, Helene Kaufler, was going to Milan. Kaufler was a history student at Zurich University. She was eager to meet Einstein's parents, and they, in turn, did not know that she was a friend of Mileva's. Einstein and Mileva soon hatched a plan. They asked Helene to question Pauline about what she thought of Mileva. It would, of course, have to be done discreetly. Helene agreed and told them she would report to them as soon as she got back.

Mileva and Einstein anxiously awaited the report. Although we don't know for sure what was said when Helene visited them, we do have a letter that Mileva wrote to her shortly thereafter, and from it we can make a pretty good guess.[12]

Helene began with, "Your mother was very nice to me, and I also met your father. He was very handsome." Einstein apparently beamed with delight upon hearing that she liked his mother and thought his father was handsome. But this was the end of the good news. Mileva had told Helene to tell her the truth even if it hurt, and as much as Helene hated to, she did just that. She told Mileva that Pauline thought she was not good enough for her son, and she was disappointed that he was involved with someone who was so much older than he. Furthermore, she put the blame on Mileva.

In the letter Mileva later wrote to Helene, she asked, "Do you think that she does not like me at all? Did she really make fun of me?" She continued, "You know, I seemed to myself so wretched now, so thoroughly wretched."[13]

Helene admitted that she joined in the "roasting" of Mileva, and apologized for it. Einstein was, of course, disappointed, but the effect on Mileva was much more devastating. She was thrown into despair. Einstein tried to comfort her, but it did little good.

THE FINAL EXAMS

Einstein and Mileva studied for the final exams together. One of the requirements of graduation was a thesis, which they were given three months to write. Einstein and Mileva selected the same topic, namely heat conduction, which had been one of Weber's areas of research and was still of considerable interest to him. Einstein later admitted that his thesis was of no consequence, and there was nothing of importance in it. But it did cause him some frustration. He wrote the thesis up a few days before the exams, but he did not use regulation paper, and when he handed it in, Weber forced him to rewrite it on regulation paper. This took up valuable study time just before the exams, and it gives us some indication of the deteriorating relations between Einstein and Weber.

Einstein and Mileva both went into the final exams in a less than desirable frame of mind. Mileva was still devastated over Pauline's reaction to their impending marriage, and she had not done well on the intermediates, which she had taken less than a year earlier. Einstein was annoyed that Weber had not covered Maxwell's theory and that the basic ideas of theoretical physics had been ignored by both Weber and Minkowski; he was also worried about his mother's reaction to his marriage. Unlike the intermediate exams, these exams would be a mixture of written and orals.

Einstein had placed first in the intermediates, but he didn't repeat the performance in the finals. All three mathematicians beat him; only Mileva

Fig. 12: Maja Einstein at age sixteen

scored lower than he. Einstein and Mileva did equally well in experimental physics, both scoring 10; in theoretical physics Einstein scored 10 to her 9. Where Mileva really fell down was in mathematics: she got 5 to Einstein's 11 in theory of functions, and she also scored poorly in astronomy, getting only 4. For his thesis Einstein received 18, whereas Mileva received 16. After a complicated weighting of the individual scores, the final grades were determined out of a maximum of 6. Einstein got 4.91, considerably less than his intermediate score of 5.7.[14] Mileva was at the bottom with 4.0. The three mathematicians ranged from 5.14 to 5.45.

Einstein was no doubt disappointed in his grade, but he passed. Mileva did not, and she was crushed. Einstein tried to console her by telling her that she could take them again the following year, and she soon decided that she would.

Looking at Mileva's grades in detail, it is easy to see that mathematics was her downfall, as it had been on the intermediates. She scored about the same as Einstein in physics, although she scored considerably lower in astronomy. Part of the problem likely was her frame of mind. She had gone through a great deal and lacked confidence; furthermore, she was very uncomfortable before the all-male examination committee in the orals. According to Einstein, she was very intelligent, but lacked "ease of understanding."[15] She also did not have Einstein's passion for knowledge and understanding.

THE "BLOWUP"

Pauline and Maja were vacationing at Melchtal, a small resort town south of Lake Lucerne, and Einstein joined them a couple of days after he completed his finals. Everything appeared to be okay on the surface when he met his mother; as usual, he was smothered with kisses. But as they traveled to the hotel, Maja pulled him aside. "Don't say anything to Mamma about the Dollie affair," she pleaded.[16] She told Einstein that Pauline was strongly against the liaison and to "please spare her feelings."

Einstein made no promises. Indeed, he now knew he had to get things out in the open, and shortly after they got to the hotel, he visited her in her room. He told her about his exam results and that Dollie had failed. Trying to appear nonchalant, Pauline innocently asked, "What will become of her now?"

Einstein knew what her feelings were, and he could feel the tension in the air, but apparently he didn't care. He was prepared for a fight. "My wife," he said curtly.[17]

Upon hearing this, Pauline threw herself on the bed, buried her head in the pillow, and began to weep. Einstein had expected a negative reaction, but he was taken back by the intensity of her reaction. After a few minutes she regained her composure and looked up at him. "You are ruining your future and destroying your opportunities. No decent family will have her," she said.

She buried herself in the pillow and began sobbing again. After a while she raised her head. "If she gets pregnant, you'll really be in a mess," she cried.

Einstein became indignant, scolding her and insisting that they weren't living in sin. His mother continued to bombard him, and he soon grew tired of it. He turned to leave when suddenly there was a knock on the door. Opening it, Einstein found one of his mother's friends, Frau Bär. She was one of the few who Einstein liked.

When Frau Bär entered the room, the atmosphere suddenly changed. Both Einstein and Pauline acted as if nothing had happened. They were all soon talking about the weather, other people at the resort, and their unruly children. It was almost time for dinner, and Pauline invited Frau

Bär to dine with them. After the meal Einstein played his violin, and when he went back to his room, it was as if nothing had happened. But he knew it was far from over, and indeed his mother soon came to say good night to him, and her bombardment continued.

"You can't marry her, it would be a big mistake."

"She's a book, like you. You need a wife."

"She's so much older than you. By the time you're thirty, she'll be an old witch."[18]

Einstein soon convinced her, however, that he had not been intimate with Mileva, and this seemed to help. Pauline must have assumed that if this hadn't happened, there was still hope. She was sure she would be able to do something to stop it.

Perhaps the most surprising part of this episode is that Einstein immediately wrote to Mileva describing everything in detail, almost as if it were a twelve-round boxing match. This would not have helped Mileva's already wretched feelings.

Chapter 6

Family Ties

Einstein stayed at Melchtal with his mother and sister through the end of July and into August. His mother's resistance to his marriage worried him, and he couldn't understand why she didn't like Mileva. She had never even met her. He finally began to wish he had kept his mouth shut. But strangely, since her initial outburst, she had said little, and he began to convince himself that she was coming around and had resigned herself to the inevitable. He was still as determined as ever to marry Mileva, and clearly the barrage had not deterred him; if anything, it had strengthened his resolve. He knew his mother was stubborn, but he assured Mileva that she wasn't nearly as stubborn as he was. He was proud that he hadn't given an inch, and he was determined to overcome any objections she presented.

In the days after the outburst, however, he did everything he could to placate his mother. He played the violin for her friends again and again, and her faced glowed as they expressed their appreciation. It was obvious that she was reveling in his popularity; it reflected on her and made her feel good. Einstein was happy to do it because it kept her mind off the "Dollie Affair," as she called it.

Einstein knew that his father would back up his mother in her opposition to the marriage, but he was sure his opposition would not be as great. By now she had no doubt written to his father, and he was expecting a letter from him. A few days later it came. Hermann pointed out that he was in no position to get married and should wait until he had a good job and some money saved. A wife was a luxury that could only be afforded by the rich. To Einstein, this put a wife and a prostitute on the same level, feelings he expressed to Mileva. The letter was just a sermon, as far as he was concerned, and he didn't take the advice seriously.

For the first few days after he arrived, the weather at Melchtal was terrible, and Einstein spent his time studying Kirchhoff. He marveled at Kirchhoff's ingenuity and was delighted at how much he was learning from the book. He hoped to finish it before he left Melchtal. Finally, however, the weather broke, and Einstein was eager to go hiking. He hiked several of the trails around Melchtal with Maja, and they were happy to find edelweiss scattered across the mountains. Einstein no doubt confided his troubles to Maja, but at this time he did not always get along with her.

He continued to write letters to Mileva, expressing his longing and love for her. "No one is as talented and industrious as my Dollie is to be found in this anthill of a hotel," he wrote.[1] He continued to reassure her that Mama hadn't mentioned the "delicate subject" and he was sure she had resigned herself. But in one letter he did admit: "She has given up open warfare and will probably wait to loose the big philistine guns when she is joined by papa." And indeed, that is what happened.

Aside from his mother and Mileva, though, Einstein had another problem. For the first while at Melchtal he enjoyed himself and didn't worry about getting a job, but within a few weeks he knew he would have to begin thinking about one. He was sure, however, that he would get one as an assistant to one of the professors at the polytechnic. The physics professors needed help in the labs, and the mathematics professors needed graders and help in the classroom. Einstein went through each of the professors in his mind. He dismissed Pernet immediately; he wouldn't want to work for him, and there was little chance Pernet would hire him. Weber was a better prospect, but Einstein hadn't got along well with him, and Weber didn't have a good opinion of him. Nevertheless, Einstein was sure that it was unimportant; aside from Mileva, he was the only physics major, and Weber needed help in his labs. Einstein was certain Weber would hire him.

Another possibility was Hurwitz, one of his mathematics teachers. Hurwitz had been in charge of the mathematics seminars, but Einstein had not attended them, and this worried him. He decided, nevertheless, to try Hurwitz. His other mathematics teacher, Minkowski, was a lost cause as far as Einstein was concerned, and he didn't even bother to try him. It is perhaps ironic that Minkowski is now revered as a great mathematician because of Einstein. Even though he had no use for Einstein, it was Einstein who was responsible for his fame. As we will see later, Minkowski put Einstein's special theory of relativity in a convenient four-dimensional form that eventually became the accepted form of the theory.

COMPASSION FOR FAMILY

On August 18 Einstein returned to Milan with his mother and Maja. Over the last few weeks at Melchtal, his mother had said little about the "Dollie affair," and Einstein was sure she was waiting for reinforcement from his father. Now that they were home, he expected them to roll out the big guns. And he didn't have to wait long.

By now Mileva, who was back at Kac, had developed a goiter, and news of it had got back to his parents. This reinforced their opinion of her as an unhealthy person. They told him over and over that Dollie would not be an asset as a wife. She was Serbian, had health problems, and was much too old for him. But Einstein held his ground. Their arguments fell on deaf ears.

As usual, however, Einstein wrote a blow-by-blow account of the sessions to Mileva. "My parents are very worried about my love for you. Mama often cries bitterly and I don't have a single moment of peace. My parents weep for me almost as if I was dead."[2] It was bad enough that he was putting her through the emotional wringer by telling her all this, but he overdramatized things with statements like "Oh Dollie, it's enough to drive one mad."

Einstein was as stubborn as ever, but he did not argue with his parents. He tried to avoid fighting with them, and over the next few weeks he was very respectful and meek, except on one occasion. His father asked him to tour company power plants at Cannetto and Isola della Scala. He wanted him to learn something about the administration of the business in case he had to take over in an emergency. Einstein agreed, then changed his mind, but he quickly changed it back when he saw the reaction and distress of his parents. He made his father promise, however, that they would also visit Venice.

Einstein enjoyed the trip, but he had doubts about taking over the business, even in an emergency. Things were not going well for his father, and he was far in debt. Einstein had little interest in going into the business, but he worried about it continuously, and now he was beginning to worry about his father's health. The stress had been hard on him, and it showed.

Einstein hoped there would be a reply from his inquiries about an assistantship at the polytechnic when he returned, but there was nothing. He was still optimistic about getting a job, but was beginning to worry. One of the bright spots of his time in Milan was his visits with Besso. Besso was working for the Society for the Development of the Electrical Industry in Milan, and Einstein enjoyed visiting him and his wife, Anna. He still complained to Mileva that Besso was a "wimp" and was disorganized and confused much of the time, but he always had praise for his keen mind. His suggestions were frequently helpful, and Einstein loved talking shop with him.

Despite Einstein's professed love for Mileva, his ties to his family were strong, and he had no intention of breaking them, regardless of how bad things became. His letters to Mileva are full of compassion for them. "You wouldn't believe how much I suffer when I see how much they both love me. . . . My parents are helping as much as possible; the *poor things* have been under continual aggravation and worry about the damn money. My dear Uncle Rudolf (the "Rich") has been nagging them terribly," he wrote.[3]

Einstein felt the pull of "family," and he had particularly strong feelings for his father, more so than his mother. One of the reasons may have been their quite different personalities. Pauline was extremely strong-minded whereas Hermann was easygoing and friendly to everyone. Einstein felt sorry for his father, and he worried about him. And the feelings were reciprocated; as we will see later, Hermann wrote a letter on behalf of his son that showed his deep love.

Einstein finally got a letter from Weber and was disappointed to find he had not been selected as his assistant. Weber had selected two mechanical engineers instead of him, and this created another problem. Einstein was going back to the polytechnic in less than a month and had planned on working on a thesis on thermoelectricity under Weber. This put a damper on things, and Einstein knew that the relations between him and Weber would be even more strained; nevertheless, he decided to go ahead with the project. Mileva was also planing to work on a thesis under Weber.

The only chance of getting on at the polytechnic now was with the mathematician Hurwitz. Hurwitz wasn't as resentful toward him as Weber, but there was still the problem of the seminars. He wrote to Hurwitz and was pleased when a few days later he got a reply. Hurwitz would consider him.

QUESTIONABLE PASSION FOR MILEVA

During the time Einstein was at Melchtal and Milan, he wrote numerous letters to Mileva, and each of them was filled with declarations of his love. From a distance he was, indeed, a very passionate and loving person when it came to Mileva. "Only now do I see how really madly in love with you I am." "I miss your two little arms and that glowing mouth full of tenderness and kisses."[4] And much more.

But, as the saying goes, "absence makes the heart grow fonder," and this seems to have been the case with Einstein. He built up an image of Mileva that was very difficult for her—or anyone—to live up to, and it seems that the image didn't endure after they were together. He went home at Christmas after promising her he would stay, and then he left again in March. There were, indeed, financial difficulties, but if he really "couldn't live without her" as he stated in his letters, he would have stayed with her despite the difficulties. It was easier for him to go back to his family. There's no doubt, however, that he and Mileva did cling to one another as soul mates. Einstein saw himself as an outcast with no one to turn to, and Mileva was a "savior" of sorts to him. And Mileva was certainly quite desperately in love with him.

Mileva brought her eighteen-year-old sister, Zorka, with her when she came back to Zurich in October. Einstein found Zorka stubborn and obstinate, but there is no mention of mental problems, which later plagued her. He referred to her as a "typical female," but quickly exempted Mileva from the category.

Einstein was disappointed in the small amount of money they were able to make from their tutoring. He no longer had his allowance from Aunt Julie, and Mileva only had a small amount of money from home, so money was badly needed. Furthermore, it was soon evident that he wasn't going to get an assistantship at the polytechnic. Hurwitz turned him down. In addition, he had planned on doing a thesis under Weber, and that soon fell through. We don't know exactly what happened, but it seems that Einstein's interest in theory wasn't to Weber's liking. He preferred to have his students work in the laboratory. By now Einstein was intrigued with capillarity (the rising of liquid in a small diameter tube) and was working on a paper he hoped to publish.

Einstein soon gave up on Weber and went to Alfred Kleiner at the University of Zurich, which was nearby, and to his delight Kleiner agreed to accept him as a doctoral student. Einstein was now spending much of his time studying Boltzmann, and he was impressed. Boltzmann was at the University of Leipzig and had made important contributions to thermodynamics and statistics; furthermore, he had also worked on capillarity. In December Einstein sent his first paper, which was on capillarity, to *Allanen der Physik.* Within a short time he got a reply that it was accepted and would be published in March. He was overjoyed; now he would be able to include a reprint of it when he applied for jobs. He was sure it would make a difference; he was a published scientist now, and they would have to pay more attention. Furthermore, he was overflowing with ideas and was confident that several more papers would be forthcoming.

He was surprised to get a letter from his mother just before Christmas pleading with him to come home for Christmas. She argued that it wouldn't look good if he stayed in Zurich with Dollie over the holidays. A detective was still checking on him in regard to citizenship, and if he found out, there might be problems. Einstein dismissed the argument, knowing that there was little chance of his being denied citizenship for having a girlfriend. After all, they weren't living together, so there was nothing immoral. Nevertheless, he decided to go home for Christmas, and he dreaded having to tell Mileva, since he had promised her he would stay. As expected, Mileva was very disappointed when he told her. More than anything, she worried that his parents would poison his mind against her while he was there.

Einstein spent Christmas in Milan and was back in Zurich in early January. His parents apparently said little about Mileva while he was home. After settling back in Zurich, he continued working on a thesis under Kleiner and began working on a second paper on capillarity. In February he got a letter from the immigration office. The detective's report was complete, and he was to come in for another interview. On February 21 Einstein was granted citizenship, but it cost him most of his savings. He had been saving for this for years, however, and felt it was money well-spent.[5]

He was delighted to become a Swiss citizen, and cherished the citizenship throughout his life, even after he became an American citizen many years later. His draft notice came a few days before his twenty-second birthday, in March. He had hated the army when he was in Munich, and one of the reasons he had left Germany was to avoid serving in it, but now he was looking forward to his military obligation. To his surprise, however, he failed the medical exam; according to the report, he had varicose veins and flat feet, and would not be up to the rigors of the marching that was required. The report stated that he was five feet seven and a half, and had a thirty-four-and-a-half-inch chest.

Einstein was disappointed, but he soon realized it was for the best in that he would be able to continue with his thesis and research without interruption. But his job prospects continued to frustrate him, and the money he was bringing in from tutoring was barely enough to exist on.

He finally decided to return to Milan; at least he would get decent meals there. He had hoped to finish his thesis before he left, but things had not been going well, and there was little chance of that. Mileva was devastated again; she was continuing to work under Weber and had not been getting along with him, so it was a difficult time for her.

MORE REJECTIONS

Einstein arrived back in Milan on March 23. He now had reprints of his article on capillarity and hoped they would be helpful in getting a job. Earlier he had sent a copy of his paper to Boltzman; now he sent a reprint to Friedrich Ostwald, a physical chemist at the University of Leipzig. He told Ostwald that his work on capillarity had inspired him, and asked for a job. But the flattery did no good; Ostwald did not reply. Subsequent letters went to Göttingen, Stuttgart, Venice, Bologna, and Pisa. The one to Göttingen went to Eduard Riecke. Einstein thought he had a good chance because Riecke had advertised for two assistants. But again he was disappointed when he received a rejection. If you look at the advertisement, however, it specifies that the applicant have a doctorate, and Einstein seems to have ignored this. He did not have a doctorate, and the prospects of him getting one in the near future were not great at that time.

After waiting for several weeks and not hearing from Ostwald, Einstein sent a follow-up letter, using the excuse that he may not have included his address on the first letter. But it didn't appear to have done any good; he still didn't hear from him. Unknown to Einstein, his father also wrote to Ostwald. Part of his letter is as follows:

> Dear Herr Professor Ostwald:
>
> My son, lacking in a post at present, feels deeply unhappy and each day the thought gains strength in him that his career has been derailed and he cannot find a connection any longer. He is moreover depressed at the thought that he is a burden to us, who are not well off . . . perhaps [you can] send him a few lines of encouragement, so he might regain his vitality and working vigor."[6]

Ostwald also did not reply to this letter, but to his credit he was the first one to nominate Einstein for the Nobel Prize only a few years later.

Einstein soon became convinced that Weber was the main source of his problems, and he dropped his name from his application forms as a reference. He replaced him with some of his teachers at Aarau and Munich. He also began to worry about the fact he was a Jew, and feeling that anti-Semitism was much more common in Germany than in Italy, he decided to concentrate on Italy. Although anti-Semitism in Germany was relatively high, and would become much greater a few years later, it was not high in most surrounding countries. Einstein's friend, Ehrat, was a Jew, and he had no trouble finding a job. Furthermore, his teacher, Minkowski, was a Jew. But even Mileva must have considered it a problem. In a letter to Helene, she wrote, "You know that my treasure has a sharp tongue and on top of that is a Jew."[7]

DETERMINATION IN THE MIDST OF FRUSTRATION

It might seem a little amazing that Einstein was able to concentrate on his studies and research in the midst of all his problems. His solution was to take refuge in his work. In many ways he was like an ostrich burying its head in the sand; he buried himself in his research to get away from personal problems. And his problems were a little overwhelming. Not only was he having trouble finding a job, but he was still arguing with his parents over Mileva. Furthermore, his father's business was on the verge of bankruptcy, and he was having trouble with his thesis. All in all, things were not easy for Einstein, but it is perhaps the difficulties that gave rise to his determination and resolve. He told Mileva that he had more stubbornness in his little finger than his parents had in their entire bodies; he was referring to their disapproval of his impending marriage, but it seems that this also applied to his studies. He was just as determined in the face of his failure to get a job. He clearly had strong feelings, and it was almost as if, after each rejection, he said to himself, "I'll show them what I can do, and they'll regret turning me down." And indeed, some of them likely did.

Despite Einstein's problems, his creativity was now blossoming rapidly, and his mind was full of ideas on capillarity, the ether, relative motion, radiation, and the theory of gases. And even though in several of his letters to Mileva he stated that he couldn't do anything without her by his side, this wasn't true. He continued to work on many problems while he was in Milan and she was in Zurich. One of them was a continuation of his work on capillarity. Actually, he was more interested in a new theory of molecular forces, and applying it to capillarity as an example. Years later he scoffed at his first two papers on capillarity, referring to them as insignificant beginners papers. They did, however, display considerable creativity and gave some indication of what was to come. Furthermore, his interest in the ether and relative motion continued, and in one letter he says that he spent four hours discussing them with Besso. Sometime during April he also saw Max Planck's paper in which Planck described a new and strange theory of radiation. It was a paper that was to revolutionize physics and introduce the idea of the quantum. It would, in fact, eventually lead to quantum mechanics. Einstein said that he had "mixed feelings" about the paper. He was disturbed by it, but he took it seriously, and as we will see later, he was one of the first to follow up on it. He also became intrigued with a theory of metals that was published by Paul Drude. Einstein's passion for his research shows in many of his letters. His letters to Mileva and others at this time are filled with science. And although he mixed personal and scientific information in his letters to Mileva, he almost seemed to prefer writing about science. In a letter to Grossmann, he stated, "It is a glorious feeling to perceive the unity of a complex of phenomena."

A RAY OF LIGHT

With the continual flow of rejections, there's no doubt that Einstein was depressed. But there was light at the end of the tunnel. He got a letter from Grossmann; he had told him about his troubles, but felt a little guilty about burdening him with his problems, so he didn't ask him for any help. He was delighted, however, when he read Grossmann's letter. Grossmann had talked to his father about Einstein's difficulties, and his father in turn had talked to Friedrich Haller, the director of the patent office in Bern. A position would be coming up in the patent office, and if Einstein applied for it, there was a good chance he would get it. Einstein was overjoyed. He sat down and wrote to Grossmann immediately. "I am truly touched by your loyalty and humanity which did not let you forget your old luckless friend. . . . I need hardly tell you that I would be happy to be granted such a fine field of activity, and would do everything not to disgrace your recommendation."[8]

There would be a waiting period of a few months, but at least he had a prospect. It was like a ray of light through a heavily overcast sky, and Einstein was delighted. The following day he received a letter from Jakob Rebstein of the technical high school in Winterthur. Rebstein had a military obligation from mid-May until July and was wondering if Einstein would like to take over his teaching position during this time. Einstein hadn't even applied for the position, but was delighted; he later found out that it was Ehrat and another friend who had recommended him. Sure that his luck was finally changing, Einstein wrote to Mileva telling her to come to Lake Como so they could celebrate.

Mileva agreed to go at first, then changed her mind. "I received a letter from home today and it has made me lose all desire, not only for having fun, but for life itself. Don't let it bother you though, go ahead and take the trip."[9]

Einstein couldn't believe what he was reading and wrote back immediately. "I just won't let up! You absolutely must come to see me in Como, you sweet little witch. It will cost very little of your time and will be heavenly joy for me." He went on to say that she would hardly recognize him, "as he had become so bright and cheerful."[10]

Mileva changed her mind again and met him at Como. On the first morning, they wandered down the main promenade in Como, then toured the city, looking at the sights. They stopped at a cafe and ordered coffee and a pastry, then decided to take a boat trip around Lake Como.

They boarded the boat just before lunch. It stopped at Cadenabbia, and they visited Villa Carlotta where they toured the lush gardens of the villa. "We weren't allowed to swipe even a single flower," Mileva wrote to

Helene later.[11] The following day they headed for Splügen Pass, which was at such a high altitude that the snow had not yet melted. Near the summit they rented a small sleigh, which had just enough room for two, with the driver standing on the runners at the back. While he drove, it started to snow, "but I held my sweetheart firmly in my arms under the coats and shawls with which we were covered," wrote Mileva.[12] To her delight, the driver thought they were honeymooners. The trip was a delightful experience for both of them.

AN UNPLEASANT SURPRISE

Einstein moved to Winterthur and started his new job as soon as they got back. He was surprised to find that he enjoyed teaching. He had to teach thirty hours a week, and one of the topics he had to cover was Fiedler's dreaded descriptive geometry. He visited Mileva each Sunday in Zurich. Life seemed to be much better, and Einstein was in a much cheerier frame of mind. But more difficulties were to come.

He got a letter from his father asking if he could help pay for Maja's education in Aarau. He requested thirty francs a month. Einstein knew this would be difficult to save, but he agreed to it. Hermann mentioned that the business was on the verge of bankruptcy, and indeed within a short time, he declared bankruptcy. Einstein felt bad, but he had been expecting it. Nevertheless, he continued to worry about his father's health.

Then on one of the Sundays that Einstein visited Zurich, Mileva had some news for him. "I'm pregnant," she said. Pauline's fears had come to pass.

Chapter 7

More Difficulties

Einstein was surprised when Mileva told him the news, but he had to have worried about the possibility. The big anxiety, of course, was that his mother would find out; he had assured her they had not been intimate. He also must have worried about the complications it would create in relation to his impending job at the patent office. But the first thing he had to do was comfort Mileva and tell her that there would be no problems, and he did. He assured her that they would get married, but she would have to be patient; he would have to find a steady job first.

They both knew that they had to keep the pregnancy secret from his parents. Furthermore, her exams were only a couple of months away, and there was the worry that the complications of pregnancy such as morning sickness would make it difficult to study.

Surprisingly, over the next few weeks Mileva's pregnancy occupied Einstein's thoughts less and less. As far as he was concerned, they would get married as soon as he got a job, and according to Grossmann he was a "shoe-in" for the job at the patent office. Still, it hadn't been advertised in the paper yet, and even when it was, it would be several months before a decision was made. His job at Winterthur would be up in July, so he had to continue sending out application forms.

HAPPINESS AND DISMAY

It's a little ironic that in his first letter to Mileva after learning about her pregnancy, Einstein started off by saying how happy he was, but it wasn't happiness about her condition. It was happiness about a paper he had just read.[1] The German scientist Philipp Lenard had just published an article describing some experiments he had performed on the generation of cathode rays by ultraviolet light. He was able to produce cathode rays (which are actually electrons) by shining ultraviolet light on a metal, and of particular importance, the brightness or intensity of the ultraviolet light had no effect on the energy of the emitted electrons; for a given wavelength of ultraviolet light, the electrons always came out with the same speed. He found, however, that he could change their speed by varying the wavelength of the ultraviolet light. This was quite surprising, and to Einstein it was an important breakthrough. Indeed, only a few years later, Einstein would explain it in a paper that would bring about a major change in physics. In 1921 he would be awarded the Nobel Prize for this work.

All that, however, was far off in the future. But the letter shows that research was still the main focus of Einstein's mind. Only later in the letter did he mention the pregnancy. It is also another indication of the enormous enthusiasm he felt for his work. In it he stated, "Under the influence of this beautiful piece I am filled with such *happiness and joy* that I absolutely must share some of it with you."[2]

Einstein was also absorbed at this time with Drude's theory of metals. It appears that he was developing his own theory when he saw Drude's theory. Drude assumed that the electrons formed a sort of "electron gas" within the metal. Einstein was impressed with the way Drude set up his theory and referred to him as a "brilliant man." He did, however, see two flaws in the theory. We're not sure what the flaws were in his eyes, but it seems he was disturbed by Drude's assumption that there were two types of charges within the metal: positive and negative. Experiments had shown only negative charges. His other objection is a little more subtle, and it was likely directed at both Drude and Ludwig Boltzmann. Drude had used Boltzmann's statistical theory in setting up his theory, but earlier Einstein had found what he referred to as a "gap" in Boltzmann's theory. Einstein was sure that both objections were obvious and that Drude would agree with him. "He will hardly be able to offer a reasonable refutation, as my objections are very straight forward," he wrote to Mileva.[3] In his letter to Drude, Einstein mentioned he was without a job, and he hoped Drude would offer him something.

Over the next few weeks he waited anxiously for Drude's reply. Indeed, he probably spent more time thinking about it than he did about Mileva's pregnancy. Finally, returning to his boardinghouse late one evening in July, Einstein saw a letter waiting for him, and it was from Drude. To his disappointment, however, Drude was not pleased with having the shortcomings of his theory pointed out. He dismissed Einstein's criticism, saying, "an eminent colleague agrees with my theory, and that's good enough for me."[4]

The reply sent Einstein into a rage. He immediately wrote to both Mileva and Jost Winteler about his outrage. "It is such manifest proof of the wretchedness of its author," he wrote to Mileva.[5] (He apparently sent the letter to her.) He went on to say that he would attack Drude mercilessly in the journal. That would, of course, have been a little difficult since Drude was the editor of *Annalen der Physik* and all submitted articles went through him. Einstein also told Mileva that he would put his scientific career on hold and look for a position immediately. He didn't care how modest the job was, he'd take it, and as soon as he got such a position, they would get married.[6]

Einstein had, for many years, seen "scientific authority" as one of the greatest enemies of truth, and he became even more convinced of it now. He saw himself as an outsider pitted against the "stodgy" scientific establishment. As it turned out, however, his threats were hollow. He never did write a criticism against Drude, nor anyone else whom he threatened. His rage only went so far.

Mileva was pleased when she read his letter, but she knew he was upset by Drude's rejection; he had his pride, and she didn't want him to accept a job that was below his abilities, knowing he would eventually resent her for it. She wrote back, "You shouldn't take a really poor job, darling; that would make me feel terrible and I couldn't live with it."[7]

COURAGE UNDER FIRE

Einstein's letters were the only bright spots in Mileva's life now. Her exams were only two weeks away, and she must have known that she wasn't well prepared, particularly for the mathematics part of the exam. Furthermore, she was still worried about what Pauline would say and do if they got married without telling her. In addition, the work on her thesis under Weber was not going well. She resented Weber because of his treatment of Einstein, and hardly a day would go by without an argument with him. She was surely beginning to wonder if she would ever complete a thesis.

July 15 was Einstein's last day of employment at Winterthur. Although he continued to write to Mileva about how much he missed her and how much he longed to be with her, rather than go to Zurich to be with her, he joined his mother and sister, who were at the Hotel Paradise in Mettmenstetten. Mileva was in a bad way and could have done with his support, but family ties again got in the way.

During his stay at Mettmenstetten, Einstein became quite disturbed because Maja was so cool toward him. She had been living with the Wintelers and had no doubt talked with Marie many times about him, and it was obvious that Maja didn't like the way Einstein had treated Marie. She must have wondered if he was treating Mileva in the same way.

Mileva took her exams in late July. Five others took the exams with her—all males. When it was all over a few days later, it must have seemed like a bad dream. Her grades were a repeat of the year before. She got an average of 4.0, and again she failed; yet everyone else passed.[8] She must have wondered how she was going to face her parents: she had failed again, and she was pregnant to top it off. Einstein encouraged her to try again, but she decided against it. Furthermore, she decided to give up on her doctoral thesis and vowed never to work for Weber again.

She was severely depressed as she got ready to go home. Worried about the reaction of her parents, she asked Einstein to write her father a letter, telling him about the pregnancy and stating that he would marry her as soon as possible. Einstein agreed. Mileva was so worried about how he would word the letter that she asked to see it before he mailed it.

As expected, her homecoming wasn't a joyous occasion. Her parents had received Einstein's letter by the time she arrived, and they were not happy. Marija, Mileva's mother, threatened to "thrash the young man" if she got her hands on him.[9] Mileva probably got more support from her father, Milos, who always looked upon her as his favorite. She had to tell them that she had also failed her exams, and that no doubt made things worse.

The worst, however, was yet to come. Shortly after she arrived home, they got a letter with a Milan postmark. Pauline Einstein had not given up her fight; fearing that she was having little influence on her son, she tried another approach. She wrote to Mileva's parents.[10] We don't know exactly what was in the letter, but it must have been very unpleasant, because it sent Mileva into a state of depression and outraged her parents. Pauline referred to Mileva as a wicked older woman who was leading her son astray, and she expressed her strong feelings against the marriage. Even though it was strongly worded, it's unlikely that she knew about the pregnancy at the time; if she had, it might have been even worse.

Mileva's parents became extremely upset and vowed that they would never accept Einstein as a son-in-law. Milos forbade Mileva from marrying him, despite her pregnancy. Now, to Mileva's dismay, both her parents and Einstein's were against the marriage. Fortunately, Mileva's parents eventually came around and began to accept the possibility of Einstein as a son-in-law.

Einstein knew nothing of this until somewhat later; Mileva did, however, write her friend Helene. She began by saying that she had ripped up several letters to her because they were so full of bitterness. She had hoped to break down Pauline's defenses, but with Pauline's acrimonious letter to her parents, she had given up. In her letter to Helene, she said, "[My] misery was due to the charming behavior of my dear mother-in-law! That lady seems to have made it her life's goal to embitter as much as possible not only my life but also that of her son. Oh Helene, I wouldn't have thought it possible that there would exist such heartless and outright wicked people!"[11]

A NEW JOB

Einstein was now back in Winterthur; he had kept his room there and was continuing his search for another job. He saw an advertisement for one in Schaffhausen, a small village about fifteen miles north of Winterthur on the German boundary with Switzerland. It was the hometown of a friend from his schooldays at Aarau, Conrad Habicht. Using Habicht as a reference, Einstein applied for the job and got it. He would be tutoring one student, Louis Cahen, an Englishman who had failed his *Matura* because of a weakness in mathematics. The pay, which was 150 francs a month, was less than Einstein had hoped for, but he would also get room and board.

Einstein did not like the school director, Jakob Nüesch, almost from the beginning. He had a gruff manner and had been strongly influenced by the militarism and strict discipline of German teaching just across the border, the very thing Einstein had hated when he was in Germany. He also soon resented the fact that he had to eat meals with the Nüesch family (there were four children in the family). On the other hand, he quickly took to his student, Louis Cahen, who wanted to study architecture at the polytechnic in Zurich once he passed his *Matura*. Cahen, in turn, enjoyed Einstein's humor and easygoing, casual way of teaching, and when Einstein discovered that Cahen felt the same way he did about Nüesch, they soon became allies against him.

Soon after Einstein started work, he received a letter from Mileva. She told him about the letter from his mother and the problems she was now having with her own family because of it. She wanted to get away—to be with him. She worried, however, that if she went to Schaffhausen, his parents might discover her there, so she told him she would stay in a village about twelve miles away, called Stein am Rhein.[12]

Einstein promised he would travel to Stein am Rhein on his days off, and judging by her letters, he wasn't as eager to see her as he had professed to be earlier. She scolded him in several of her letters.[13] He wrote to her that he was too broke to come on one occasion. She offered to send him money. On another occasion she wrote, "Now, you're not coming tomorrow again! and you don't even say: 'I'll be coming on Sunday instead.' . . . You know, if you don't come at all, I may just leave. If you only knew how terribly homesick I am you would surely come."[14] She stayed until the end of November, then returned to her family. She was now seven months pregnant.

Fig. 13: Alfred Kleiner

Einstein worked on his doctoral thesis, which was on molecular forces, while he was in Schaffhausen, and on November 23 he submitted it to Dr. Kleiner at the University of Zurich. At this point he did not have a high opinion of Kleiner, and in one of his letters to Mileva he "dared" Kleiner to reject his thesis. "[If he does] I'll publish his rejection along with my paper and make a fool of him. . . . It's really terrible, all the things these old Philistines put in the path of people who aren't of their ilk."[15] But privately he was worried. Although the thesis wasn't on the theory of conduction in metals, Einstein had somehow worked Drude's theory into it, and he castigated both Drude and Boltzmann in the process. He worried about Kleiner's reaction.

Soon after this, Einstein learned that Cahen's mother was paying Nüesch 4,000 francs a year for his schooling, and Nüesch was only paying him 150 francs a month. Einstein now began looking upon Nüesch as both a cheat and a scoundrel. He finally came to a decision: If Nüesch was getting that much money, he could pay more. Einstein therefore told him that he would like to take his meals at the local inn and to be reimbursed for them, instead of eating with his family. Nüesch was not happy but said he would talk it over with his wife. He later told Einstein that they had decided against it. "You knew what the conditions were when you came here, and there is no reason to deviate from them," he said. Einstein replied, "Good, as you wish. I'll have to give in for the time being."[16] He then implied that he would be looking for another job. Given that he had had considerable trouble finding *any* job, he was obviously tempting fate. But it worked. Nüesch agreed to provide his meals at the local inn, much to Einstein's relief. He knew that Nüesch was now furious with him, but he didn't care.

The following evening Einstein ate his last meal at the Nüesch's. He found a letter propped up against his plate; opening it, he found it was from Grossmann. By now he had almost given up on the job at the patent office in Bern. Several months had passed, and there had been no advertisement of a position. When he read Grossmann's letter, he was delighted to find that an advertisement for the position would appear within the next few days. A day later the ad appeared in the *Federal Gazette*, and there was no doubt that Haller had tailored it to Einstein's qualifications. It specified a university education of "mechanical-technical, or specifically physical direction," the latter words implying a physicist. Until then, the patent office had never hired a physicist.

With the Bern position now almost certain, but several months away, Einstein began hatching a plan. When he moved to Bern, he would take Cahen with him. Cahen disliked Nüesch as much as he did and was eager to go along with the plan. Einstein would continue tutoring him in Bern, but now the 4,000 francs that Nüesch was getting would go to him. Einstein was, in essence, stealing Nüesch's student, and he relished the thought. "That wretched old fellow will be stunned when I tell him. He's such a dreadful scoundrel," he wrote to Mileva.[17] Louis wrote to his mother, hoping to convince her to allow him to go to Bern. But his mother wasn't enthusiastic about the move. Her husband had just been admitted to an insane asylum, and she wasn't up to making decisions. She preferred things left as they were. Einstein was disappointed but thought it was probably for the best.

His troubles with Nüesch, however, weren't over. Nüesch soon became upset with Einstein's teaching methods. He felt he was being too lax and there was no discipline in his class; in short Einstein was too friendly with his student. This was not the proper way to teach, according to Nüesch, and he reprimanded Einstein for it. Einstein was not happy, but he kept his mouth shut, at least initially.

VISIT TO KLEINER

For several weeks Einstein had been waiting for some word from Kleiner about his thesis. Finally, on December 19 he decided to go to Zurich and find out what was going on. Einstein's thesis was on molecular forces and the kinetic theory of gases, but he was also continuing to work on other problems. The problem of electrodynamics of moving bodies had occupied his mind for several years. To him it appeared to center on relative motion, but strangely, few scientists had thought about the importance of relative motion in relation to electric and magnetic fields. Einstein was sure it was the answer to the enigmatic relation between the two fields, and to deal with it he would have to look closely at the concept of absolute motion, in other words, motion that appears the same to everyone in the universe. Newton had postulated that absolute motion existed, but the concept didn't make any sense to Einstein, and it seriously bothered him.

When Einstein visited Kleiner, he was dismayed to find that Kleiner hadn't even begun to read his thesis.[18] Kleiner promised, however, to read it over the Christmas break. Trying to be polite, Einstein implied that there was no rush. The conversation then turned to Einstein's work on the electrodynamics of moving bodies. He outlined his ideas to Kleiner and told him that he had also devised an experiment to check on them. Kleiner was impressed and encouraged him to publish. Einstein decided to take his advice, but it actually took three years for him to do so. By then he had modified his ideas considerably. They eventually became his special theory of relativity.

Einstein also talked to Kleiner about his job situation, and Kleiner offered to write him letters of recommendation whenever he wanted them. Einstein left his office in high spirits. Earlier Kleiner had been shortsighted and stupid in Einstein's eyes, but that had suddenly changed. "He's not quite as stupid as I thought, and moreover, he's a good fellow," he wrote to Mileva.[19] He was now sure that Kleiner would accept his thesis. But as it turned out, he was in for a surprise.

Einstein decided to stay in Schaffhausen to work on his paper on the electrodynamics of moving bodies over the Christmas holidays. He did, however, take Christmas day off to spend with his sister at Hotel Paradise in Mettmenstetten. He was still not on the best terms with her, and he reported later that the visit was not a success. He had kept Mileva's pregnancy secret from all his friends, and it appears that he also kept it secret from Maja. It is not known when his mother heard of it, if she ever did.

Although Einstein and Maja didn't always see eye to eye, she did have considerable sympathy for Mileva, and it came out in a visit home sometime after the new year. Pauline began her usually tirade against Mileva, but Maja had apparently heard enough and began to stick up for Mileva. It seems that she also told her to leave Einstein and Mileva alone, and let them get married. Pauline, as you might expect, exploded, and Maja quickly left the house. For some time the two were not speaking to one another. Shortly after this, Pauline wrote, "This Miss Marić is causing me the bitterest hours of my life. If it were in my power, I would make every possible effort to banish her from our horizon. I really dislike her."[20]

OLD FRIENDS

On December 27 Einstein met with his friends Grossmann and Ehrat. Grossmann was now teaching at a secondary school in Frauenfeld. Einstein had applied for the same job and had written to Grossmann congratulating him after he was selected over him. Grossmann was working on a thesis on non-Euclidean geometry under Wilhelm Fiedler. The title of his thesis was "On the Metrical Characteristics of Co-Linear Functions." Einstein had no idea what it meant or was about. In fact, he felt sorry for Grossmann, having to work under Fiedler. Ironically, his general theory of relativity, which came several years later, centered on non-Euclidean geometry. Ehrat had apparently gone to Minkowski for a thesis topic, but the topic he was given seemed so silly that he quickly switched to Geiser. Neither man had completed his thesis, and Einstein was proud that he had. "See, your Johnnie finished his paper first, despite being hounded in the process," he wrote to Mileva.[21] But as it turned out, he was a little overconfident. Grossmann received his doctorate the following year in 1902; Ehrat, on the other hand, had to wait until 1906, and as we'll see, Einstein also had a long wait.

LIESERL

Mileva was in Novi Sad. Her parents had bought a winter home in the city to get away from the farm in Kac. She was now in her ninth month and was expecting the baby any day. The birth, when it came, was an extremely difficult one for her, and left her weak and exhausted. She was so weak, in fact, that Milos had to write to Einstein telling him of the birth.

Einstein got quite a shock when he received the letter from Milos. He hadn't heard from Mileva in quite a while, and knew that she was due, so he was worried. When the letter came from Milos, rather than Mileva, he expected the worst, since death of the mother during childbirth was still quite common at that time. Einstein said he was "scared out of his wits" and was afraid to open the letter.[22] It was a considerable relief to him when he found out she was weak but still alive.

Einstein had hoped for a boy, Mileva for a girl, and it turned out to be a girl. She was named Lieserl. They had kept Mileva's pregnancy secret from everyone except perhaps Helene; none of Einstein's friends knew about it. Einstein expressed considerable interest in the baby at first, writing such things as: Who does she most resemble? Is she healthy? He even urged Mileva to send him a picture of her. There had been talk earlier of putting her up for adoption, but there is also some indication that Einstein wanted to keep her. In one of his letters, he said, "The only problem that still needs to be resolved is how to keep Lieserl with us; I wouldn't want to have to give her up."[23] He went on to tell her to ask her Papa what should be done.

But it appears that Einstein never set his eyes on her. When Mileva returned to Switzerland several months later, Lieserl was not with her. Indeed, the secret of her birth was so closely guarded that no one knew of her throughout Einstein's life. Einstein never mentioned her. It was only after both Einstein and Mileva were dead and the cache of love letters was unveiled that her existence came to light. It's therefore natural to ask: What became of her? As it turns out, no one knows for sure (except, of course, Einstein, Mileva, her parents, and perhaps Helene). It appears that Mileva's parents looked after her for the first while. About a year after Lieserl's birth, in 1903, Mileva returned home to discover that Lieserl had scarlet fever, and she wrote a letter to Einstein, who expressed some concern. After that, she just disappeared. Several people have searched the birth records in and around Kac and Novi Sad, but there is no record of her. It seems likely that she was put up for adoption, or perhaps adopted by a nearby relative. Ironically, many years later, when Einstein was quite old, a woman came forward claiming to be Einstein's daughter. Einstein immediately hired a detective to check her out, and it turned out that she was a fraud; she was much too old to be Lieserl. So the mystery remains.

Having to give up Lieserl appears to have had a serious effect on Mileva. It left her saddened, moody, and remorseful. Einstein's son, Hans, who knew nothing about having an older sister, remarked in later life that his mother spent a lot of time brooding over something that she referred to as "very personal." She was encouraged to get it out in the open, but she couldn't; she kept it secret until she died.

After Lieserl's birth, Einstein's troubles weren't over. He had moved from Nüesch's house and now had his own room, but his relationship with Nüesch hadn't improved. Einstein and Cahen were still talking about the possibility of going to Bern together. It's not known whether or not Nüesch found out about the plan, but he called Einstein into his office one day, and a violent argument ensued, with Nüesch finally telling him that he was fired. Einstein could no longer restrain himself, and he apparently told Nüesch off in no uncertain terms, then stormed out of the school, never to return.

The only thing Einstein could do now was to go to Bern and wait for the job at the patent office. There was still the lingering worry that he might not get the job, but he tried to be confident. Before he left he decided to visit Kleiner at the University of Zurich to see if he had finished reading his thesis yet. And indeed, Kleiner had read it, but there were problems. Einstein had worried about his criticism of Drude and Boltzmann, and this is what Kleiner zeroed in on. It seems that he found the challenge to Boltzmann particularly repugnant and told Einstein that the thesis would have to be rewritten. It's known that he didn't reject it outright, however; otherwise Einstein wouldn't have got his entire thesis fee of 230 francs back, and we know he did. It appears that Kleiner told him to withdraw it. Einstein likely agreed quite quickly, since the 230 francs he had posted must have seemed like a lot of money at that point. It would get him through two or three months in Bern. He applied for the return of the fee on February 1, 1902.

Einstein was disappointed when he left Kleiner's office. Another disappointment to add to his long list. He had been looking forward to getting his doctorate, knowing that it would be of considerable help in getting a job. Only a few days earlier he had been proud of himself for finishing his thesis before Grossmann and Ehrat. He must have wondered how long his streak of bad luck would last.

Chapter 8

Gaining New Insights

Einstein arrived in Bern on February 4, 1902, and soon found a one-room apartment at 32 Gerechtigkeitsgasse, near the Nydegg Bridge. He had enough money for a few months, but knew it would likely be at least that long before he started at the patent office, if indeed he got the job, and there was no guarantee of that. He was depressed with the way things had gone. He hadn't completed his doctorate, and almost two years had passed since he had graduated, and he still didn't have a steady job.

Still, Einstein wasn't going to leave his fate to chance. Bern was a college town, and many people needed tutors. He therefore put an ad in the Bern paper for a math and physics tutor. He planned to charge two francs per lesson and offered to give free trial lessons. Writing to Mileva a few days later, he mentioned that two people were interested, an engineer and an architect. It's not known if either of them actually became his students, but he did manage to land Louis Chavan, a French-speaking Swiss technician. Chavan kept meticulous notes and also kept a record of what he thought about his teacher. He wrote, "His striking brown eyes radiate deeply and softly. His voice is attractive, like the vibrant note of a cello."[1]

OLD FRIENDS

Einstein loved to wander along the arcade (covered sidewalks) in Bern. Bern is, in fact, famous for its arcade. It was on one of these strolls that he met a classmate from Aarau, Hans Frösch, who was studying medicine at the University of Bern. Frösch invited Einstein to come along with him to a lecture on forensic pathology. Knowing that this was something that might interest Mileva, Einstein decided to go. The lecture was on mentally deranged criminals, and the lecturer brought two people on the stage to demonstrate his ideas. One was a female pyromaniac who set fires when she got drunk, and the other was a pathologic swindler. Einstein was so fascinated that he decided to go to the meeting every week; he wrote to Mileva about it as soon as he got home. He also told her that he had visited with his old friend Conrad Habicht, and was thoroughly enjoying Bern.

A few days later he got a letter back from her. She was annoyed that he was having such a joyous time with his friends and asked if he had forgotten her and his new daughter. Einstein wrote back immediately trying to console her. "Don't be jealous of Habicht and Frösch," he said, "what are they compared to you! I long for you every day—but I don't show it because it's not 'manly.' But I'd certainly rather be with you in a provincial backwater than without you in Bern."[2] Einstein went on to tell her about someone he had met who was working at the patent office. He had told him that the work was boring, but he said he wasn't put off. "Certain people find everything boring," he wrote, "I'm sure I'll like it and will be grateful to Haller for as long as I live."

A VISITOR

Einstein had only been in Bern a short time when he got a visit from Max Talmud. Talmud had brought books to Einstein when he was a teenager and had encouraged him, sure that he had a great future ahead of him. Einstein was no doubt surprised by his visit, since he hadn't seen him in about eight years. Talmud had dropped in on Einstein's parents in Milan and was concerned when they seemed reluctant to talk about him. Sensing that something was wrong, he asked for Einstein's address and went to Bern to visit him. After looking around Einstein's apartment and talking to him for a few minutes, he was sure he knew what the problem was. The apartment was small and the furniture old; furthermore, everything reeked of stale tobacco. It had to be quite a letdown after his parents' large, beautiful house in Munich and their large apartment in Milan. Talmud was sure that the apparent estrangement between Einstein and his parents was because he had not lived up to their expectations; he had obtained a degree from the polytechnic but was now jobless and appeared to be living in poverty. As they talked, Einstein complained about the difficulties he had had in trying to obtain a job and a doctorate. He said that people were putting obstacles in his way because they were jealous of him. He had to have been thinking of Weber, Drude, and perhaps Kleiner, who had rejected his thesis. The major problem, of course, was that Einstein still had a chip on his shoulder and a smoldering resentment against authority. He hadn't got over the idea that he was an outcast, and was still as determined as ever to show the scientific world what he could do.

SOLOVINE

Einstein did not have a lot of luck with his first ad in the Bern paper, so at Easter he placed another one. This time he decided to charge three francs for his lessons. Maurice Solovine, a philosophy and physics major at the University of Bern, saw the advertisement and decided to investigate. Over the past few months, he had become dissatisfied with the education he was getting. He was interested in philosophy, but the few classes in philosophy that he had taken at the University of Bern had disappointed him. It seemed to him that philosophy was befuddled; it seemed to lack coherence and direction, and he wondered if it was the way it had been presented. He had also taken physics classes, mainly from the department head, Aimé Forester. Forester's major interests were

Fig. 14: Maurice Solovine at the time of the Olympic Academy

astronomy and meteorology, and, to Solovine, his knowledge of advanced physics was limited. He found his lectures lacked depth and were very elementary, and as a result they were disappointing to him. He wanted more.

Solovine found his way to Einstein's rooming house and climbed the stairs to his apartment. Einstein invited him in. Solovine's first impression was positive: he was struck by the clarity and sparkle of Einstein's large brown eyes. He told him that he was a Jew from Romania and had been studying philosophy but wasn't satisfied with what he was getting at the University of Bern. He must have been pleased when Einstein told him that he had also become disenchanted with philosophy. Solovine went on to tell Einstein that he also had a serious interest in physics. He complained about Forester's lectures, describing them as superficial, and he was annoyed that he never explained anything. In one lecture, he said, Forester discussed the radiative properties of radium, but said nothing about why it had these properties or what was significant about them. This had turned Solovine off.

Einstein's reaction to Solovine's complaints had to be encouraging. He knew what Solovine was talking about and sympathized with him. Physics was, indeed, frequently taught in the way Solovine described, and whenever Einstein had encountered it, he had been as discouraged as Solovine was. This was not for people who really cared about physics; people who had a real feeling for physics wanted more. Einstein had always had a deep respect for the mysteries of nature, and he felt that if you studied them until you thoroughly understood them, you would appreciate them much more.

The two men talked for over two hours, and then Einstein walked him down to the street where they talked for another half hour before they parted. They agreed to meet the following day.

Einstein began lecturing to Solovine a few days later, but within a short time the two men decided that formal lectures were too restrictive and stifling. Free discussion would serve their purposes much better, with Solovine asking as many questions as he wished and expressing his opinion whenever he wanted. A few weeks after they began, they were joined by Einstein's friend, Conrad Habicht, who was now a math major at the University of Bern. Habicht had recommended Einstein for the job with Nüesch, and no doubt got an earful about Nüesch when they first met in Bern. Habicht was the son of a bank director in Schaffhausen; he had studied math and physics earlier at the Universities of Zurich, Munich, and Berlin, and was now working on his doctoral thesis at the University of Bern.

Fig. 15: Conrad Habicht at the time of the Olympic Academy

The three men would usually begin their discussion sessions with a meal of sausage, cheese, fruit, and tea or coffee. After the meal they would study books by well-known philosophers and scientists. Among them were Karl Pearson, Ernst Mach, Henri Poincaré, John Stuart Mill, and David Hume. A strong camaraderie soon developed between them. Solovine and Habicht were thoroughly impressed with Einstein and his seemingly effortless ability to explain things so clearly. "He explained things in a slow and even voice, but in a remarkably clear way. To make his abstract ideas more easily grasped, he would sometimes take examples from everyday experiences," said Solovine.[3] Einstein, in turn, was pleased with the enthusiasm and conscientiousness of his students.

Despite the joy he experienced in the company of Solovine and Habicht, by early spring Einstein was encountering financial problems, and he didn't know how much longer he could hang on. The small amount he was bringing in from tutoring was hardly enough to keep him in tobacco, and his savings were nearly depleted. Mileva was writing regularly and occasionally sending him food packages, but he wasn't sure how much longer he could survive.

Solovine and Habicht could see that he was having problems. Realizing that his birthday was coming up, they decided to treat him to caviar.[4] Earlier, Solovine had pointed out the delicacy to Einstein in a food shop, and he had expressed interest in it, saying he had never tasted it. They were sure it would be a big surprise, and were eager to see his reaction. On his birthday the meeting was at his apartment. Solovine put the caviar on three plates and placed one in front of each of them. Einstein had been thinking about the principle of inertia prior to the meeting and had come to an interesting conclusion in relation to it; he was anxious to tell the three men about it, and soon he was absorbed in a discussion of it. He mouthed down forkful after forkful of the caviar without looking at it while he talked. Finally, when his plate was clean, Solovine said, "Do you realize what you have been eating?" Einstein looked down at his plate in surprise. "No, what was it?" "Caviar," replied Solovine. "So that was caviar," said Einstein in surprise. Then after a short silence he laughed and added, "It serves you right. . . . If you offer gourmet food to a peasant like me, you know it won't be appreciated."[5]

On another occasion Solovine saw a poster for a violin concert that interested him. He suggested that they forgo the next meeting and go to the concert which conflicted with the meeting, but Einstein vetoed the suggestion. As much as he loved music, he felt they should continue their study of Hume, since it was now at a critical point. On the night of the concert, Solovine succumbed and bought a ticket. The meeting was to be at his place that night, so he told his landlady to tell Habicht and Einstein that he had been called out of town on important business. Leaving four hard-boiled eggs for them, he wrote a note, "To my dear friends, some hard-boiled eggs and a salutation." Einstein and Habicht knew immediately what the "urgent business" was, and they decided to teach him a lesson. Both men knew that Solovine hated tobacco smoke, so Einstein took out his pipe and Habicht lit up a big cigar. They puffed until the apartment was full of smoke, then to make sure he got the message, they piled all his furniture on the bed, stacking it almost to the ceiling. They then left a note, "To a dear friend, thick smoke and a salutation."

Solovine said that when he got home he had to open every window in the apartment to get rid of the smell. Everything including his blankets, pillow, and bedclothes were saturated with the smell of smoke, and it was almost morning by the time he finally got to sleep.[6]

THE PATENT OFFICE

By late May, Einstein had almost given up on the patent office; then finally to his delight, he got a letter from it. The director, Freidrich Haller, wanted him to come in for an interview. Einstein was overjoyed. On the following day he went down to the office, which was in the new Postal and Telegraph Administration building on Genfergasse. The patent office was on the third floor. Haller greeted him and took him into his office. Over the next couple of hours he bombarded Einstein with questions about his expertise and experience. Many of the inventions that were being submitted to the patent office were now of an electrical nature—things such as alternating current devices, polyphase current machines, telephone technology, and devices associated with wireless technology—and most of the examiners were mechanical engineers with limited experience in electricity and magnetism. Haller hoped that Einstein would be able to fill in the gap.

Einstein had, indeed, encountered many electrical devices in Weber's lab and was an expert in the theory of electromagnetic waves. Nevertheless, his experience was limited, and Haller no doubt sensed this, but he was not disappointed. Einstein had considerable knowledge in other areas such as heat and thermodynamics, which would be useful in relation to inventions in the refrigeration and heating areas. Einstein also had another advantage, which he likely didn't mention to Haller. He had been exposed to the inventions and patent applications of his Uncle Jakob, who was an avid inventor. He therefore had a feeling for the inventors themselves.

The interview must have gone well because a couple of weeks later, on June 16, Einstein received a letter stating that he had been "provisionally elected Technical Expert III Class at an annual salary of 3500 francs."[7] The money was double what he would have earned from an assistantship at a university, so he was delighted. Still, it was not a large amount for a married man with a family.

On June 23 Einstein reported for work. He was not familiar with the procedures of the office, and Haller had to spend considerable time with him over the first few weeks. He showed him how to analyze an invention, how to check to see that it wasn't infringing on an existent invention, and how to judge its usefulness. Einstein was critical by nature; he had found fault with Drude's and other theories, and was critical about many thing in physics, so it wasn't a new experience for him. He found the job to be an interesting challenge and was pleased that it required more thought than he had anticipated. Each invention was different, and each required a different approach. Although Einstein was a theorist, he had considerable experience in the lab, and Weber's lab was as up-to-date as any lab in Switzerland.

Einstein was taken aback at first by Haller's gruff manner and coarse language, but he soon took it in stride and laughed at it. His main deficiency at first was a lack of familiarity with technical drawings and how to interpret them, but Haller helped him overcome this. Haller found Einstein to be an enthusiastic and fast learner, and Einstein, in turn, held Haller in high esteem; he later referred to his "good brain and excellent character."[8]

THE OLYMPIC ACADEMY

Einstein continued his discussion sessions with Solovine and Habicht after he began work at the patent office. They soon decided that their group needed a name and picked the rather pretentious sounding "Olympic Academy." It may have been a joke, but Einstein later said that it was a less childish academy than some of the respected ones he got to know after he became famous. There's no doubt that the three men enjoyed each other's company, exchanging good humor and fun, but there was also serious work. They studied books by well-known philosophers and scientists, then discussed the ideas in them. They went through Karl Pearson's *Grammar of Science*; Mach's two books, *Mechanics and its Development* and *Analysis and Perception*; along with John Stuart Mill's *Logic*. Einstein was particularly enthralled by Henri Poincaré's book, *Science and Hypothesis*, because it went into considerable detail on the concepts of absolute and relative space and time. He also wondered about David Hume's philosophical idea that there was no pure logical path from observation to the scientific laws that explained them. He had doubts about it.

Although the meetings were held alternately at the three men's apartments, on many occasions their discussion sessions would consist of a hike. Thun, a picturesque village on a lake about twenty miles away, was a favorite weekend destination. They also occasionally climbed a mountain near Bern, called Mt. Gurten; on one occasion they hiked to the peak at midnight and didn't get back until morning. At the peak, with the stars shining brightly overhead, they discussed cosmology and the wonders of the universe. Their arguments and discussions were quite animated and boisterous at times, and they surely had a great influence on Einstein. Throughout his life he needed a "sounding board" for his ideas, and Solovine and Habicht were excellent in this respect. Einstein explained his ideas on molecular forces, statistical physics, relative and absolute motion, and so on to them, and in the process, he sharpened his insights. Habicht and Solovine asked questions that made Einstein think more deeply about the subjects, and in the process many things became clearer to him.

It's well known to all scientists that merely explaining something in detail to another person, particularly a problem that they have been working on, frequently results in the solution. This had to have been the case with Einstein and his two students. Although Einstein had now given up on his quest for an academic position, his passion for physics was as

Fig. 16: Einstein in 1905, the year he published his special theory of relativity. Courtesy of Lotte Jacobi.

great as ever, but he now felt that the key to academic success was publication. If he could get his ideas out to the scientific world via publications, someone would eventually see their merit. Einstein was still, indeed, hard at work on his research. After working for some time on a theory of molecular forces and on the kinetic theory of gases, he came to the conclusion that there was a shortcoming in the foundations of thermodynamics and in the way large ensembles of molecules were treated. In a gas, for example, each of the molecules had a certain set of properties (i.e., velocity, momentum, energy) at any given time, but you couldn't deal with individual molecules. Somehow you had to determine the "average" properties, and this is what you had to use in calculations. He also wanted to derive the second law of thermodynamics from mechanical principles. This was the so-called gap he had found in Boltzmann's theory earlier. Einstein felt that he had to overcome it before he could proceed to problems in which particle statistics were needed. In a series of three papers, Einstein looked into this problem. The first was titled "Kinetic Theory of Thermal Equilibrium and the Second Law of Thermodynamics." It was submitted in June 1902 and published in September.[9] The paper opens with the statement that Einstein intends to fill in the "gap" in the foundations of heat equilibrium and the second law of thermodynamics by using simple mechanical principles and probability theory.

The second paper in the series came in January 1903; it was published in April.[10] It followed up on the work of the first paper and was titled "A Theory of the Foundations of Thermodynamics." The third paper titled, "On the General Molecular Theory of Heat,"[11] came in 1904. These three papers became classics, and they showed how much Einstein had matured intellectually; in particular, they gave some insight into his tremendous potential. They were the foundations of a new science, which we now refer to as statistical mechanics. Unfortunately, they were not new. Unknown to Einstein, Willard Gibbs in the United States had published the same material just before he did. Einstein had limited access to physics journals and books while he was in Bern. In addition, Gibbs's articles had been written in English, and Einstein did not read English. Gibbs published his book, *Elementary Principles in Statistical*

Mechanics, in 1902, but Einstein did not see the book until it was translated into German in 1905. It had to be a disappointing blow for him; still, it showed what he was capable of.

A DEATH IN THE FAMILY

Things were looking good for Einstein after he got a job at the patent office. He enjoyed the work, and it left him considerable time for research; furthermore, he also had two excellent students with whom he had a good relationship. He still didn't have a new topic for a doctoral thesis, but there were many good prospects. In addition, it seems that shortly after he started at the patent office, Mileva returned to Zurich, and he was visiting her. In a letter dated June 28, he signed off with, "Farewell my little sweetheart; we'll meet Monday at 6:00 at the little tower."[12] There was still no mention of marriage, however. His parents—particularly Pauline—were still strongly against it.

In early October, Einstein got a message from Milan: "Please come home immediately." It had to have been something he had been worrying about and dreaded, since he knew his father was in poor health. When he arrived, he found that his father had, indeed, suffered a heart attack. The years of pressure from his business had caught up with him. Einstein was informed that there was no hope and that his father would likely die. The news devastated him.

Hermann was only fifty-five, but had suffered through three business failures and was now deep in debt, mostly to Rudolf Einstein, the husband of Pauline's sister Fanny. Einstein had always been very close to his father and blamed Rudolf for his death. For months Rudolf had been pressuring him for repayment of the money that he had borrowed.

Hermann died on October 10. Just before passing away, he called Einstein in and gave him permission to marry Mileva. After saying good-bye to everyone, Hermann asked everyone to leave; he wanted to be alone in his last minutes. For years Einstein felt guilty about not being with him at the end; according to several of his friends, he felt a "shattering sense of loss."

He later referred to his father's death as the "deepest shock he had ever experienced."[13] At one point he said to Maja, "Why couldn't it have been me."

Pauline was left with a large debt, and she now had no way of supporting herself. She asked Einstein to help. Although he now had a decent salary, it was no doubt a burden to him, but he agreed to help her. What was particularly annoying to him, however, was that Pauline was going to live with his hated Uncle Rudolf. Einstein was distressed that his mother was going to live with the man whom he felt had caused his father's death.

Nothing is known about Pauline's reaction when she heard that Hermann had given Einstein permission to marry Mileva, but she was in no shape to fight it, and finally also gave her permission somewhat reluctantly. Nevertheless, she never accepted Mileva. Pauline stayed with Rudolf and Fanny in Hechingen for about ten years, then moved to Berlin for a short while. She became the housekeeper for a widower in Heilbronn, Germany, and in 1914 she managed the affairs of Hermann's brother, Jakob, who was now widowed. Maja had graduated at Aarau and had passed the entrance exams for the University of Bern, but she would now have to wait to attend university. She took a job in Trieste.

Chapter 9

A Passion for Understanding Nature

Einstein returned to Bern after his father's funeral. He now had his family's permission to marry Mileva, but he didn't rush into it. The reason for this was twofold: he was still depressed over his father's death and it took some time to get over it, and he also now had some reservations about getting married. In later years he said that he was unsure of himself at his point. It had been six years since he had met Mileva, and he now knew her fairly well. He knew that she was moody and somewhat possessive, and had a temper; on the other hand, he felt a sense of duty and knew he should do the right thing. She had gone through a lot for him: she had borne his child, she had endured his parent's scorn, and there's no doubt she was expecting him to marry her.

And indeed they were married about three months later. Announcements were placed in the Bern, Zurich, and Novi Sad newspapers, but aside from that, there wasn't much fanfare, and the marriage ceremony was a simple affair. It took place on January 6, 1903, at the courthouse in Bern. No one from either family attended. This may seem a little strange, but Pauline certainly wasn't likely to come, and there were still strong reservations against the marriage from Mileva's parents, mostly because of Pauline's letter to them. They hadn't fully accepted Einstein as a potential son-in-law. One person who could have come was Maja, but she didn't. They needed witnesses, of course, and Habicht and Solovine gladly volunteered.

After the ceremony Einstein and Mileva had their photograph taken, and then they went to a restaurant with Solovine and Habicht for a meal and to celebrate. It was late by the time Einstein and Mileva got back to his apartment at 18 Tillierstrasse, and true to form, Einstein had either forgotten or misplaced his keys and had to get the landlady out of bed to let them in.

From all indications, the first few years of the marriage was relatively happy. Shortly after he got married, Einstein wrote to Besso, "Well, now I'm an honorably married man, and leading a very nice, comfortable life with my wife. She takes care of everything exceptionally well, cooks well, and is always cheerful."[1] For her part, Mileva wrote to Helene, "I am, if possible, even more attached to my dear treasure than I already was in the Zurich days. He is my only companion and society and I am happiest when he is beside me."[2]

Einstein hoped that married life would free him from domestic duties and give him more time to pursue his research. For the first while this may have been true, but once they had children, it became more difficult for him. The Olympic Academy went on as usual; the only difference was that it now had one more member, namely Mileva. Mileva was, by nature, very shy and reserved, but she liked Solovine and Habicht, and became close friends with them. She particularly liked Solovine, who went out of his way to be nice to her. According to Solovine, however, she wasn't a particularly active member of the academy. He said she listened quietly to the arguments and discussions, but never joined in, even when they got quite heated. For the most part, she was a silent observer, but of course she may have discussed things with Einstein after Solovine and Habicht left.

When the discussions got particularly loud, the three men usually took to the streets, but Mileva never came along. Also, on many occasions they went on hikes as they discussed various problems. Mileva never accompanied them on these hikes.

Although Mileva still had some interest in physics at this stage, it seems that her passion for it did not come close to that of Einstein and the other members of the academy. Her failure to get a diploma at the polytechnic had no doubt cooled her ardor.

OTHER FRIENDS

Solovine and Habicht were not Einstein's only friends in Bern. At the patent office he had met Josef Sauter. Sauter was eight years older than Einstein and had studied engineering at the polytechnic; indeed, he had at one time worked as Heinrich Weber's assistant. He was now a Swiss citizen, but had been born in France. Although he was an engineer, he was interested in physics and had studied Maxwell, Boltzmann, and other physicists on his own, and he liked to talk to Einstein about what he had learned. He was also a member of the Bern Natural Society and invited Einstein to come along to one of the meetings. The society was made up of professors, high school teachers, medical doctors, and other professionals, and talks were given at each of the meetings. The topic of these talks ranged considerably, from biology and medicine to physics and chemistry.

Through Sauter, Einstein also met Paul Gruner, a high school teacher in Bern who was a privatdozent in physics at the University of Bern. This was of particular interest to Einstein since a privatdozent was the first step to becoming a professor. Privatdozents could teach, but received no salary; they were allowed, however, to collect fees from their students. Becoming a privatdozent was not easy; a candidate had to go through what was called "habilitation," which required both a thesis and a trial lecture. Furthermore, the applicant had to have a doctorate, and although Einstein had submitted a doctoral thesis to Kleiner, he still didn't have his doctorate. To Einstein's delight, however, Gruner pointed out that there were "special circumstances" at the University of Bern, which didn't exist at most universities.[3] The doctorate and habilitation could both be dispensed with if there were "outstanding achievements." Einstein had now published four papers and was in the process of publishing a fifth. Furthermore, all had been published in the prestigious journal *Annalen der*

Physik, which in itself was a considerable achievement. Einstein was sure that these papers would be recognized as an outstanding achievement.

The major problem, of course, was that the committee that selected the privatdozents had to recognize the importance of the papers. Einstein's papers, particularly the ones on thermodynamics, were at the frontiers of research, and they were not easy to understand. Even worse, the final decision was made by Aimé Forester, who had little interest in theoretical work. As might be expected, in almost record time, Einstein's proposal was turned down; the committee clearly wasn't impressed with his achievements or his papers. Einstein lost his temper when he heard of the outcome. "The university is a pigsty," he wrote. "I wouldn't lecture there anyway. It would be a waste of my time."[4]

Despite the setback, Einstein kept busy. He was working on a third paper on thermodynamics and was attending the meetings of the Bern Natural Society regularly. In December he presented his first talk at the meetings, which was titled "On the Theory of Electromagnetic Waves." Also, in July he had saved up enough money for a belated honeymoon, and he and Mileva went to Lausanne, on Lake Geneva, for several days.

MILEVA'S VISIT HOME

Sometime over the summer Mileva made a trip back to Novi Sad. It appears that the main purpose of the trip was to put Lieserl up for adoption. There is some evidence that Mileva took her to an agency in Belgrade.[5] The final parting with Lieserl was particularly hard on Mileva; she became increasingly gloomy after it. Furthermore, during the trip she discovered that she was pregnant again, and she worried that Einstein might not approve. With some trepidation, she wrote to him about her condition. Einstein wrote back, "I'm not the least bit angry that poor Dollie is hatching a new chick. In fact, I'm happy about it."[6] He knew that she was depressed at having to give up Lieserl, and he hoped that a new child would take her mind off it. He also told her to get back as soon as possible. Three and a half weeks had passed and "a good little wife shouldn't leave her husband alone any longer." In November 1903 the Einsteins moved to 49 Kramgrasse, close to Bern's famous clock tower. Their new apartment had two rooms and was reached by a steep staircase. One of the rooms had a large window that overlooked the street below. The apartment is now a museum. Their first son, Hans Albert, was born there on May 14, 1904. With the birth of his son, Einstein's life changed; he now had to help with the baby-sitting, but this didn't keep him from his work. He usually had a pad in one hand and a pencil in the other, while he was rocking the crib.

Hans remembers his father as quite attentive when he was young. He said he made him a little cable car out of match boxes. "It was one of the nicest toys I had at the time," said Hans later.[7] Hans also remembers that his father was always trying to educate him beyond what he received at school, and he also tried to instill in him a love of music.

THE FAILED COLLABORATION

Einstein wrote frequently in his early letters to Mileva about a scientific collaboration after they were married. As it turned out, this didn't happen. According to Hans, "It was hard to understand, because she originally had studied with him, and had been a scientist herself. But somehow or other, with the marriage she gave up practically all her [scientific] ambitions."[8] Although she participated in the Olympic Academy, she contributed almost nothing to the discussions and seemed to have little interest. Much of her lack of enthusiasm probably goes back to her double failure of the final exams at the polytechnic and her failure to complete a doctoral thesis under Weber. Also, if you look at the letters between them, Einstein's are filled with physics. And although many of Mileva's letters are lost, the ones that remain seem to indicate that she had only a moderate interest in physics. There is almost no science in her letters, with the exception of discussing her upcoming exams, or a brief discussion of a lecture she attended. She does not reply to Einstein's enthusiastic discussions about physics, and it seems that this remained true after they were married. Einstein, in fact, complained about her lack of enthusiasm and turned to others to fulfill his need. Solovine and Habicht were his main sounding boards at this time, but he also spent a lot of time talking with Sauter and Gruner.

John Stachel of the Center for Einstein Studies at Boston University gives three possible reasons for the failure of the collaboration:[9]

1) Mileva's talents in physics were modest.
2) She lost the inner self-confidence and drive necessary to pursue a career in physics.
3) Einstein failed to encourage her.

Let's begin with the first of these. Although her grades in physics at the polytechnic were close to Einstein's, there is no indication she was truly talented or creative. Aside from her thesis, she never worked on an independent project, and she seemed to have difficulties with mathematics. Her mathematical grades were always her lowest; in fact, her overall failure at the polytechnic was due primarily to her weakness in mathematics.

One of the problems was that Mileva was very shy and insecure, and she didn't have a lot of self-confidence. It certainly doesn't appear that she had enough self-confidence to break out on her own, independent of Einstein. She appeared to have little interest in science while married to Einstein, and even less in later life. At that time, of course, it was very difficult for a woman. She saw the obstacles that Einstein was having early on, and she knew that he was more talented. She must have known, therefore, how much more difficult it would have been for her. There were, of course, several husband-wife teams who enjoyed collaborational success, such as Marie and Pierre Curie, and Paul Ehrenfest and his wife, and they may have had some influence on her.

It seems, however, that Einstein sensed that she had little enthusiasm soon after he married her, and as a result, he didn't encourage her. According to friends, she was very reserved and said little, even to her family. Einstein no doubt soon realized she would never be a good "sounding board," but he had several people who were. He didn't need her, so he ignored her scientifically and let her go about her household duties, which seemed to be okay with her. Eventually, in fact, Mileva began to resent science, and to some degree, his fame; she felt they were coming between her and her husband. To her, he was spending too much time with science. This is evident in her letters to Helene. In one of them she writes, "You see, with such fame, not much time remains for his wife. I read a certain maliciousness between the lines when you wrote that I must be jealous of science, but what can one do, the pearls are given to one, to the other the case."[10]

A PASSION FOR RESEARCH

This was a critical time in Einstein's life. He was in his mid-twenties, and it is well known (and was well known then) that this is the age when people are the most creative. Among scientists there is the saying that "once you're over thirty, it's too late." Granted, a few people have done their most creative work after thirty. Erwin Schrödinger, for example, was forty when he formulated quantum mechanics, and as we will see, Einstein did some of his greatest work when he was over thirty. But for the most part, truly creative work in science is generally done at an early age. Of course, many scientists still do excellent research in later life, but it is generally not as creative as their earlier work.

Einstein's creativity was peaking at this time, but his passion had begun many years earlier. Einstein himself credits his early "wonder" about the things around him for giving him his thirst for understanding nature. In his autobiography he writes of seeing this "wonder" for the first time when he was a child of four or five. He was sick, and his father gave him a compass. That the needle behaved in such a strange way "made a deep and lasting impression on me," he said.[11] He soon came to the conclusion that there was "hidden meaning" behind many things that people never bothered to think about.

Einstein referred to a second "wonder" of a different nature that occurred when he received a geometry book when he was about twelve.[12] He was extremely pleased with himself after he was able to prove the theorems and exercises in it. He said that most people soon lost their wonder of the things around them when they got older, but he never did.

We get a feeling for his passion for science through the many quotations that are accredited to him. One of the most famous is: "The most beautiful and most powerful emotion we can express is the sensation of the mystical." In a similar vein, he is attributed with saying: "The fairest thing we can experience is the mysterious. It is the fundamental emotion that stands at the cradle of true art and true science. He who does not know it and can no longer wonder, no longer feels amazement, is as good as dead, a snuffed-out candle."[13]

These two quotes provide us with a good idea of Einstein's tremendous feeling and passion for understanding nature and his awe for the "workings" of the universe. To him, anyone who could stand and look at the stars in the night sky and not feel a sense of its majesty was as good as dead.

Two other statements that accentuates the feelings of joy and comfort he found in science are: "The most incomprehensible thing about the universe is that it is comprehensible,"[14] and, "When I have no special problem to occupy my mind I love to reconstruct proofs of mathematical and physical theories that have long been known to me. There is no special goal in this, merely an opportunity to indulge in the pleasant occupation of thinking."[15]

But, like any scientist, he knew that the struggle for understanding nature was not easy. He wrote, "The years of anxiously searching in the dark for a truth that one feels but cannot express, the intense desire and alternatives of confidence and misgiving until one achieves clarity and understanding, can be understood only by those that had experience[d] them."[16]

And despite his success he was still a modest man. "I have no special talent. I am only passionately curious."[17] But he gives a simple explanation of his genius in the following: "The ordinary adult never gives a thought to space and time problems. . . . I, on the contrary, developed so slowly that I did not begin to wonder about space and time until I was an adult. I then delved more deeply into the problem than any other adult or child would have done."[18]

Einstein wrote about his thinking process at many times throughout his life. Two quotes related to it are as follows:

> A new idea comes suddenly and in a rather intuitive way. That means it is not reached by conscious logical conclusion. But thinking through it after you can always discover the reasons which have lead you unconsciously to your guess.[19]

> The words or the language, as they are written or spoken, do not seem to play any role in my mechanism of thought.[20]

Einstein had a tremendous power of concentration, even at a young age. Maja remarked about it in her memoir. It is also illustrated in an interesting anecdote by Banesh Hoffmann. Hoffmann and Leopold Infeld worked as collaborators with Einstein at Princeton. According to Hoffmann, whenever they got particularly hung-up on a problem, they would go to Einstein and explain it to him. After hearing it, he would say, "I will think a little," and he would develop a faraway stare as he thought about the problem. Then a few moments later he would come up with the solution. Hoffmann said he got annoyed at himself in several cases for not thinking of the solution.

During the years after his marriage, Einstein had many distractions. According to one visitor, wet clothes were strung across the kitchen drying when he saw him; the room smelled of diapers and stale smoke, and puffs of smoke arose every so often from the stove, but these things didn't seem to bother Einstein. He had the baby on one knee and a pad on the other, and every so often he would write an equation on the pad, then quickly rock the baby a little faster as he began to fuss.

Einstein was thankful that the work at the patent office was not too demanding and left him time to occasionally pull out his pad at work. He would, of course, have to put it away quickly if Haller came along, and there is no indication that he was ever reprimanded for this.

The Olympic Academy was still meeting regularly, but to Einstein's disappointment, Habicht was offered a job to teach in Schiers in the fall of 1903 and announced that he would soon be leaving. Then a little later, Solovine left, too. Einstein, as you might expect, kept in touch with both men through the mail. For a while Habicht's younger brother, Paul, attended, but eventually the meetings came to an end.

Einstein began thinking of Michele Besso, who was now in Trieste. It had been a struggle for him over the past few months; he had gone into business on his own as a consulting engineer, and he wasn't a good businessman. Passing the bulletin board at the patent office one day, Einstein noticed an announcement for a new position; it was for a Technical Expert Class II and carried a salary of 4,800 francs a year. He was still Class III and was making 3,600 francs a year. The job appealed to him, but he also thought of Besso and wrote to him about it. If Besso was working at the office with him, it would be an excellent opportunity to discuss his work with him.

Einstein and Besso both applied for the position along with about thirty others. To Einstein's delight, Besso was selected. Haller gave Einstein a consolation prize in that he raised his wage to 3,900 francs a year, but he didn't raise him to Class II. When Besso moved to Bern, Einstein made sure he got an apartment very close to his so that they could walk to work and back each day with one another. The two families soon became inseparable. Usually reserved, Mileva appeared to like Besso's wife, Anna, and they visited back and forth regularly. Einstein enjoyed Besso's enthusiasm and depth of thought, and over the following years, Besso became an excellent sounding board for Einstein's ideas.

MILEVA'S CONTRIBUTION

Einstein was married to Mileva during his most creative years, particularly the years just before 1905 (one of his most productive years), and since she was also a physicist, it's natural to wonder if she played any role in his breakthroughs. A number of articles have, in fact, appeared that claim her role was much greater than previously assumed.

Evan Harris Walker of the Walker Cancer Research Institute of Edgewood, Maryland, published an article in *Physics Today* titled "Did Einstein Espouse His Spouse's Ideas?"[21] In the article he suggests that Einstein took credit for the basic ideas of his theory of relativity when they really came from Mileva. He even went as far as implying that Mileva wrote his doctoral thesis, and also suggested that Einstein tried to cover up everything. His justification is that Einstein referred to his research at this time as "our research" in several of his letters to Mileva. Looking closer, we see that these letters were actually written many years before his breakthroughs of 1905, so it's not a very strong argument.

Dr. Senta Troemel-Ploetz, a psychotherapist, also published an article on the subject, titled "Mileva Marić, the Woman Who Did Einstein's Mathematics." And a biography of Mileva by Desanka Trbuhovic-Gjuric also suggests (without any evidence) that Mileva did more than she is credited with. John Stachel of the Center for Einstein Studies at Boston University takes all of these people to task in his book *Einstein from B to Z*.[22] He presents several very strong arguments against the suggestions. He points out that Einstein likely discussed his discoveries with Mileva, but she made no claim to any of his ideas or discoveries during her lifetime, and in her writing she always referred to "his" research and "his" papers. The claims that she was involved came from others.

The suggestion that Mileva was in any way responsible for Einstein's theories would be hard to prove. The letters between Einstein and Mileva give a strong indication of Einstein's passion for science, in that they are filled with science, whereas Mileva's are not. Walker claims, however, that the letters in which she discusses science were purposely destroyed. Stachel states in his article that this is very unlikely. He goes on to point out that Mileva never did any research on her own, and never published anything, as Marie Curie did, for example, after her husband, Pierre, died.

It also seems strange that anyone would suggest that Mileva did Einstein's mathematics (as Troemel-Ploetz does) when it was well known that she was relatively weak in mathematics. According to her letters to Einstein, it was her major worry in both the intermediate and the final exams, and it was mainly mathematics that caused her to fail both times in the finals. On the other hand, both Solovine and Habicht mention that they were in awe of Einstein's agility with mathematics and the ease with which he used it, even though he told them he was slightly distrustful of it. Einstein admitted, however, that he had shortcomings in mathematics at the time, but what he meant by this is that he was not well-rounded; in other words, there were many branches of mathematics that he was not familiar with. It does not mean that he was uncertain of himself in basic math, and it is basic mathematics (mainly algebra and simple calculus) that he used in most of his early papers. As Stachel mentions, if Mileva had to help him with it, his career in theoretical physics wouldn't have lasted long.

It wouldn't be fair to say that Mileva made no contribution. She apparently read through his papers of 1905 and some of the earlier ones before he sent them in for publication, and she was a strong supporter of his work at this time, at least judging from her letters to Helene. Furthermore, it seems likely that Einstein talked to her to some degree about physics, although as we will see, it is clear that his major sounding board just prior to 1905 was Besso. Einstein stated several times that he valued Besso's insight, whereas he stated that Mileva lacked "ease of understanding." Also, in the years before Besso moved to Bern, Solovine and Habicht were his major sounding boards.

Einstein thanked Besso for his contribution in his paper on relativity, but he did not thank Mileva or anyone else, yet there's no doubt that others helped him. He obviously felt that Besso's contributions was much greater than anyone else's. The fact that he didn't reference anyone else is perhaps not surprising since the paper contains no references whatsoever at the end—something that is almost unknown in modern papers. Not even Newton or Galileo were acknowledged or mentioned for their earlier work.

A LETTER TO HABICHT

In May 1905 Einstein sent a letter to Conrad Habicht, describing four papers that he was in the process of publishing.[23] He promised to send him reprints as soon as he got them. He described the first paper as "very revolutionary," saying that it dealt with radiation and the energetic properties of light. And indeed, it was very revolutionary; it literally changed the study of physics and won him the Nobel Prize in 1921. For several years, however, it was so controversial that few believed it.

The second paper was a determination of the sizes of atoms and molecules. It might be hard to believe, but several well-known scientists still didn't accept the idea of atoms and molecules. This paper helped convince the scientific world that they did, indeed, exist. The third paper explained the strange phenomenon known as Brownian movement, in which tiny inanimate bodies such as pollen vibrate or move around when placed on a liquid such as water.

Einstein didn't refer to the fourth paper as revolutionary and said it was still in a rough stage. He said it was on the electrodynamics of moving bodies that employs a "modification of the theory of space and time." This was his special theory of relativity, and it would eventually set the world of physics on fire. Yet, as late as 1920 it was still controversial, and even though it was easily worthy of the Nobel Prize, he was actually awarded the prize not for his theory of relativity, but for his paper on light.

These four papers would change the world of physics forever, and would make 1905 one of the most important years in the history of physics. Only one other year compares with it, and it was 1666, the year in which Newton formulated his laws of motion, law of gravity, theory of color and spectra, and invented calculus.

It's interesting to note that both men were in their mid-twenties and near their peak of creativity and genius. Newton was twenty-four in 1666, and Einstein was twenty-six in 1905. At this point in his life, however, Newton had published nothing; Einstein, on the other hand, had published four respectable papers.

Chapter 10

The Miracle Year: 1905

Einstein decided to try for his doctorate again. He was sure it would help him at the patent office, and it was, of course, a prerequisite for an academic career. Although such a career looked a long ways off, Einstein still hoped that one day it would come. He had two choices: the University of Bern or the University of Zurich. Neither appealed to him; he had been turned down for a privatdozent by the University of Bern, and Alfred Kleiner at the University of Zurich had rejected his earlier thesis. He decided to ask Joseph Sauter's opinion. Sauter told him to stick to the University of Zurich, and Einstein decided to follow his advice.

But what would he use as a thesis topic? He had been working in several areas: radiation and the quantum, electrodynamics and relative motion, and atoms and molecules. He was sure his work on the quantum was too controversial, but Kleiner had found his work on the electrodynamics of moving bodies interesting and had urged him to publish it. Einstein knew, however, that it wasn't complete; something was still needed to bring everything together, but he decided to give it a try. He wrote up what he had and sent it to Kleiner.

Over the next few weeks, he watched the mail for a reply. Finally it came. Opening the letter, he found that his thesis had been rejected. According to Kleiner he had had some difficulty understanding the mathematics. He wrote that he had shown it to others at the university, and they had found it incomprehensible.

Einstein was annoyed; to him it seemed like "another waste of time." But he put the experience behind him and continued working on his research. His attention in recent weeks had been directed at Max Planck's 1901 paper on heat radiation. Planck had introduced a new concept called "quanta," and with it he had been able to get a formula for the emission of radiation that appeared to fit experimental data perfectly. Earlier Wilhelm Wien and, independently, Lord Rayleigh had derived formulas for the same phenomenon, but both formulas were in conflict with experiment. Wien's formula could not account for low frequencies, and Rayleigh's formula could not account for high frequencies. Planck's formula was in agreement across the spectrum. Yet it had caused little excitement; few believed it was correct, and even Planck had reservations about it.

The problem with Planck's formula was that it was not a "classical" formula, in the usual sense. The idea he used in formulating it was strange, and as such, was hard for most scientists to accept. In his theory, radiant energy could only be emitted in discrete amounts, and this conflicted with Maxwell's theory. Planck called his little discrete bundles of energy "quanta."

When Einstein first saw Planck's paper, he was disturbed, but when he studied it in more detail, it began to make sense to him, and unlike most scientists, he took it seriously. Of most importance, however, he took an important step beyond the theory by extending it to light. He postulated that light was made up of "particles" in the same way that matter is made up of atoms. His "light particles" were discrete bundles of energy—Planck's quanta.

Einstein, as usual, began his paper by pointing out a difficulty. He stated that there was a conflict between the way physicists treated matter and light. Matter consisted of particles, whereas light was continuous. This meant that when the two theories were brought together—for example when light interacted with matter—a clash was inevitable. It was almost blasphemy to suggest that Maxwell's theory was wrong, and Einstein didn't do this. In fact, he praised Maxwell's theory, but he suggested that it might not be the whole story.

According to Einstein, light was a particle, but it still had a frequency, so it also had to be a wave. Einstein knew the suggestion was radical, and he knew he would have to give some justification for it; in other words, it had to make contact with experiment. But he was ready for this. Several years earlier he had read a paper by Philipp Lenard. Lenard had shown that electrons are released from a metal when light is shone on it. It was a curious effect that we now refer to as the *photoelectric effect*. The energy of the released electrons was independent of the intensity of the radiation, which didn't make sense in Maxwell's theory. Their energy was proportional to frequency; furthermore, there was a cutoff frequency below which no electrons were ejected.

Einstein assumed that if an electron absorbed a quantum of light (later called a *photon*) of a certain frequency, it would gain the energy of the photon. If the frequency, and therefore the energy that was absorbed, was high enough, the electron would be ejected from the metal. In the process, however, it would undergo a number of collisions before it got to the surface and would therefore lose energy. Furthermore, there had to be a minimum energy required for it to get to the surface and break free. We now refer to this as the "work function."

Einstein wrote up the paper and submitted it to *Annalen der Physik* on March 17, just a few days after his twenty-sixth birthday. It was published on June 9. He titled the paper "On a Heuristic Viewpoint Concerning the Generation and Transformation of Light."[1] He used the word "heuristic" to fend off criticism. It's interesting that Planck was the editor who would either accept or reject the paper, and he accepted it without comment, even though later on he mentioned that he didn't like the idea, or take it seriously.

ANOTHER THESIS

With the paper on the photoelectric effect in the mail, Einstein began to think about submitting another doctoral thesis. He had become so disgusted with the doctoral charade that he was tempted to forget it, but he knew he had to try again. He talked to Besso about what he should use as a topic. He knew it couldn't be controversial. He had been working on atoms and molecules, and was interested in proving their existence in a way that nobody could refute.

He poured sugar into his tea as he talked to Besso. As he stirred, he looked down at the tea. "When I put sugar into the tea, the viscosity is changed," he said to Besso.[2] He stirred the tea some more. "It seems to me that the viscosity might be related to the size of the sugar molecule, and therefore if we knew the viscosity we could determine the size of the sugar molecule." After thinking about it for a while, Einstein decided that the determination of the size of a molecule was a good thesis topic; it was not controversial, but was important. Besso agreed with him.

That evening Einstein looked into the problem in more detail. How would he set up the problem? The first thing he had to do was decide on the basic assumptions. He began by assuming that the sugar molecule was much larger than the water molecule; this seemed reasonable. He then assumed that the sugar molecule was approximately spherical. With this he was able to relate the change in viscosity (when the sugar was dissolved) to the total volume of the sugar molecules. He then considered a swarm of sugar molecules dissolved in water and found he was able to obtain an expression for the diffusion coefficient (a number representing the speed of diffusion) of the sugar, and it, in turn, depended on the size of the sugar molecule. In short, if the diffusion coefficient and viscosity of the sugar solution were known, the size of the sugar molecule could be determined.

He told Besso about his breakthrough the following morning. All he had to do was look up the diffusion coefficient and viscosity of sugar in water. There was a handbook at work that would have both numbers, and soon after getting to the patent office, Einstein made the calculation. He found that the sugar molecule had a diameter of one ten-millionth of a centimeter.

Over the next few days, Einstein wrote up the work into the form of a thesis, and in early April, he submitted it to Kleiner. He knew it would take a few weeks before he would hear anything and was surprised when he got it back within a few days. It was rejected because it was too short; it was only seventeen pages long.

Einstein had to have been annoyed, but he took it in stride. He read through the thesis carefully again. The result was so important he found it hard to believe that Kleiner had rejected it. Finding nothing wrong, he added one sentence and sent it back.

Kleiner was no doubt surprised to get it back so fast, and although it didn't appear to be much longer, he looked at it much more carefully this time. The idea and the method were both impressive, but the mathematics was a little difficult for him to check. He decided to pass it on to Heinrich Burkhardt, the head of the mathematics department. He told Burkhardt that "the arguments and calculations are among the most difficult in hydrodynamics and could only be approached by someone who possessed understanding and talent." Several days later Burkhardt returned the thesis with the comment, "What I have examined I found correct in every respect, and the manner of the treatment testifies to a thorough mastery of the mathematical methods concerned."[3] Kleiner then informed Einstein that the thesis had been accepted.

Einstein was delighted; he was now "Doctor Einstein." Over the next few days, Einstein wrote up the thesis for publication in *Annalen der Physik*.[4] To his annoyance, he got it back from Paul Drude for revisions, so it was not published until about six months later. Einstein was pleased with his work on atoms and molecules and decided to follow up on it. If a large sugar molecule was jostled around by water molecules, it seemed reasonable that something even larger might be jostled in the same way. Indeed, this "something" might be large enough to see in a microscope. Molecules could not, of course, be seen in a microscope, but perhaps their "effects" could be seen. If this was the case, you would "see" the molecules indirectly, and this would be ample proof of their existence.

BROWNIAN MOTION

Einstein imagined the sugar molecule in water being replaced by a tiny speck of matter—very small, but large enough to be seen in a microscope. At any temperature above absolute zero, the water molecules have energy and move around; in fact, the higher their temperature, the greater their movement. It seemed to Einstein that the water molecules would knock the specks of matter around. He explained the idea to Besso and asked for his opinion. Besso was sure he was talking about what was called *Brownian motion*. Einstein had never heard of it.

Brownian motion was discovered, or at least first reported in detail, by Scottish botanist Robert Brown about seventy-five years earlier. Brown had observed grains of pollen in water using a microscope and had noticed that they had a trembling motion. He later showed that the same thing occurred when finely ground spheres of glass and other materials were placed on water. The effect had been named after him, and although there was some suspicion that the motion was caused by the water molecules, there had been no detailed explanation of it.

Over the next few weeks, Einstein worked on the problem. He visualized the large spheres being knocked around by the vibrating water molecules, but he knew it was unlikely he would be able to calculate the velocity of the spheres directly. A few simple calculations soon showed him that a large sphere (compared to the water molecules) would move over a hundred times its diameter every second if unhindered. But it was hindered, and would undergo collisions, so its overall movement would be greatly changed. He couldn't calculate the detailed displacement of a given sphere, but he could calculate the "average or mean displacement" of one, and it should be measurable under a microscope. He made the calculation and determined that a tiny sphere with a diameter of one-thousandth of a millimeter would move by a one-thousandth of a millimeter in one second, or six-hundredths of a millimeter in a minute. This would be observable. Over the next few days he wrote up the paper and sent it for publication.[5]

Einstein was pleased with his results. He wasn't "seeing" molecules, but he had predicted how far molecular collisions would push a small grain of pollen or other material, and this was like seeing them indirectly. If observation bore out his predication, few would doubt the existence of molecules and atoms.

RELATIVITY

Over the past few weeks, Einstein had published three important papers, but the most important one was still to come.

Einstein's fourth paper was on a topic he had been thinking about for years—seven years, according to his recollection. He had tried to use it as a thesis, and it had been rejected, but he knew when he submitted it that it was not complete.

Einstein's earliest encounter with the problem came when he was sixteen; he had tried to visualize what it would be like to chase a light beam. He mentioned the problem several times in his letters to Mileva, and he discussed it with Kleiner after he submitted his first thesis. Einstein saw the problem as a discrepancy between dynamics and Maxwell's electrodynamics. He knew, for example, that there was a problem associated with the relation between the motion of a conductor in the form of a wire loop and a magnet. If a magnet is moved through the loop a varying current is created, which according to Maxwell's theory of electrodynamics, produces an electric field that induces a current in the wire. He found no problem with that. But if you reverse things, keeping the magnet at rest and moving the wire loop, no electric field is generated according to accepted theory. In this case, however, there is also a current. But according to convention, it is created by an electromotive force, and this electromotive force is caused by the motion of the conductor relative to the magnet.

To Einstein, this didn't make sense. It was two different explanations for the same phenomenon. The only important thing, as far as he was concerned, was the relative motion between the two objects. Yet science explained the phenomenon in two different ways, depending on which object was moved.

Einstein was sure it wasn't a problem with Maxwell's theory; he was sure it went to the foundation of physics—to the concepts of space and time. Two hundred years earlier Newton had introduced the idea of "absolute space" and "absolute time." Einstein didn't like either of them. The idea that time was the same throughout the universe, regardless of the motion and position of the observer, was repugnant to him. Both Ernst Mach and Henri Poincaré had expressed doubt about the idea, and Einstein agreed with them.

Einstein saw the problem as follows. Suppose you had two clocks (we'll call them A and B). Assume that they both show the same time. Now suppose that clock B is transported to a nearby star, and we can see it through a powerful telescope. The image we see, however, takes a finite amount of time to reach us, since it travels at the speed of light, which is finite. The time on the clock in the image is therefore quite different from that of clock A, depending on how far the star is away from us.

Also, in its trip to the star, clock B will be moving at a high speed relative to clock A. What effect will this have? Will clocks A and B click off seconds in the same way? Einstein knew that we use what is called the *Galilean transformations* to relate the times on the two clocks, and according to it, the seconds would be the same length.

Einstein and Besso discussed the problems thoroughly one evening after work, and Einstein continued thinking about them as he went to bed. He was sure that time was the key, and if he was to thoroughly understand it, he would have to consider when events in the universe were simultaneous and when they were not. He arose refreshed the next morning, and the moment he was up, he was thinking about the problem. He thought about it as he washed and shaved. Suddenly it came to him—almost like a bolt out of the blue (as breakthroughs usually do). It was so logical that he was surprised he hadn't thought of it before. Time would, indeed, be different for two observers in relative motion, and their times could be related, but not by a Galilean transformation. A new transformation would be needed, but this new transformation could be determined if certain postulates were assumed.

When Einstein saw Besso the next morning, he was jubilant, and said,

"Thank you. I have solved the problem."[6] He explained his ideas to Besso. He had decided that time was not absolute, but depended on the relative motion of the observers. Over the next few days, he derived the new transformation law; we now refer to it as the *Lorentz-Einstein transformation law*. When he applied it to the problem of the magnet and the conductor, the solution was simple, and it depended only on the relative motion of the two objects. Furthermore, it made a lot more sense than two separate explanations.

Einstein was soon writing up a paper for publication. He began with this statement:

> It is well known that Maxwell's electrodynamics—as usually understood at present—when applied to moving bodies, leads to asymmetries that do not seem to be inherent in the phenomenon.[7]

He went on to describe the problem of the magnet and the conducting wire. After setting down his basic postulates, he went on to define "simultaneity" for the reader (it means "at the same time"). He then considered the relationship between length and time, concluding the section with

> Thus we see that we cannot ascribe absolute meaning to the concept of simultaneity; instead, two events that are simultaneous when observed from some particular coordinate system can no longer be considered simultaneous when observed from a system that is moving relative to that system.

Finished with the preliminaries, Einstein derived the new transformation equations. Over the next few days, he continued working on the paper, carefully checking everything. He showed that velocities should he added in a new way, then went on to electrodynamics, the Doppler effect (a change in the frequency or wavelength of a wave due to motion of the source or observer), and aberration (in relation to starlight it is the apparent shift in a star's position due to Earth's orbital motion). The last section of the paper was directed at the dynamic properties of an accelerated electron.

Einstein's first assumption in the new theory was that all motion in the universe is relative. Relative motion does not seem important to us because we live on Earth, and all motion here is relative to Earth. But if you were out in space with nothing around, and something moved past you, there is no way you could say definitely that it was moving and you were stationary, or vice versa. The only thing you would know for certain is that there is relative motion between you and the object. According to Einstein, relative motion is the only type of motion that has any meaning.

Einstein's second postulate was that the speed of light was a constant throughout the universe, and it has some strange and interesting consequences. To imagine one of them, consider a rocketship and a large searchlight that are sitting side by side on Earth. You know the light beam from the searchlight will travel out at 186,000 miles per second; for example, if you point it in the direction of the Moon, it would take 1.3 seconds to get there, and if you pointed it in the direction of the Sun, it would take approximately 8 minutes to get to it.

Suppose now that you blast off in the rocketship, and within a short time it is traveling at 185,000 miles per second. That's not possible, of course, because of gravitational effects, but we won't worry about that for now. Next, assume that a second or so later somebody turns on the searchlight (it is still sitting on Earth). What happens? According to Einstein, the beam from the searchlight will catch up with the rocketship and pass it almost immediately. Furthermore, if you measured the velocity of the light beam, you would find that it is traveling 186,000 miles per second faster than the rocketship, despite the fact that the rocketship is going 185,000 miles per second. It's almost as if two plus two isn't four. Normally, if you have a train going 40 miles per hour and another train passes it at 60 miles per hour, the second one is going 20 miles per hour faster than the first. But near the speed of light this isn't true; actually, it's not true at all speeds, but it's more noticeable near the speed of light.

One of the most important things to come out of Einstein's theory was that the ether was superfluous. It was no longer needed, and therefore didn't exist. Light was still a wave, but it didn't need a medium to propagate it.

Einstein showed that his transformation equations gave rise to other strange effects. The best way to understand them is to consider a rocketship passing overhead. Although it has to be traveling at a very high speed, we will assume we can see what is going on in it with a telescope. Einstein's first prediction was a contraction in length. Suppose we have two meter sticks. We attach one of them to the side of the rocketship and keep the other one at our side. Then assume the rocketship passes overhead several times at speeds close to that of light. We will see that the meter stick shrinks in the direction of motion; the greater the speed, the greater the shrinkage, until at the speed of light it appears to shrink to nothing. Of course, the rocketship also shrinks in the same way.

Because it shrinks to nothing at the speed of light, the speed of light is obviously an uppermost limiting velocity. According to Einstein, nothing can achieve this speed (only light itself). Material objects can travel as close to the speed of light as you wish, but they cannot attain it. Furthermore, a clock aboard the rocketship will also appear to run slow compared to one back on Earth. And again, the faster the rocketship is traveling, the greater will be the difference. Furthermore, at the speed of light the clock on the rocketship will appear (to someone on Earth) to stop (assuming we could see it). The overall result of this is that if a rocketship took off from Earth and traveled at speeds close to that of light, then returned to Earth, the people in the rocketship would be younger than they would have been if they had stayed on Earth. A final effect occurs in relation to the mass of the object. It increases as its relative speed increases, again becoming infinite at the speed of light.

Einstein mailed his paper on relativity to *Annalen der Physik* on June 30, and it was published on September 26.[8] He was soon hard at work on a follow-up paper. In it he showed that if a body emits energy, as in the form of radiation, its mass decreases. Einstein titled the paper "Does the Inertia of a Body Depend on Its Energy Content?"[9] It was a short paper—only four pages long—but it contained an extremely important result. A year later he was able to show the complete equivalence between mass and energy. In short, the mass of a body is a measure of its energy content. The proportionality factor between them is the velocity of light squared, so there's a lot of energy tied up in a small amount of matter. It was this result that eventually led to the creation of the atomic and hydrogen bombs and nuclear reactors.

A VACATION

Einstein was so wrapped up in his work that he had little time for his family. Indeed, the intense work leading up to his relativity paper so exhausted him that he had to take to bed for several days after it was in the mail. Because of his dedication to his work, his relationship with Mileva suffered. Already gloomy and depressed after having to give up Lieserl, she continued to become more depressed as he ignored her. She was anxious for a vacation, and with some effort, she finally convinced Einstein to take some time off. She wanted to take him to Novi Sad to meet her family. Einstein wasn't anxious to meet them; he remembered that her mother had said she would thrash him if she ever got her hands on him. Nevertheless, he relented and said he would go. Mileva acted as a guide, introducing him to her parents, relatives, and friends around Novi Sad. Einstein was no doubt uncomfortable, but luckily for him, much of the attention was directed at their son, Hans.

Mileva's parents were friendly to him, but there was a language barrier, and Mileva had to act as a translator. Einstein was relieved when the day of departure finally came. He was anxious to get back to Bern to see if his paper on relativity had been accepted, and indeed it had.

Einstein waited patiently for a reaction to his papers. He expected considerable criticism of his relativity paper and also his paper on light quanta. He knew it would take time to hear anything, but over the next few months nothing happened. Finally, however, he got a letter from Max Planck asking for clarification of some points in his relativity paper. He was delighted and continued to correspond with Planck for some time after that. Soon a few other letters began to trickle in; most were requests for reprints. Heinrich Zangger of the University of Zurich wrote about his Brownian motion paper, requesting a reprint and some clarifications.

All in all, though, the lack of interest was disappointing to Einstein. Many people, including well-known scientists, had to have read his papers by now—particularly the one on relativity—and he had heard almost nothing. There appeared to be little reaction to them. He began to worry that they wouldn't help him. He still longed for an academic position but was now beginning to think that he might be stuck at the patent office for the rest of his life.

MORE FRUSTRATION

Life was so strange—so much had happened with his breakthroughs and the overwhelming feelings of joy that had come with them. Still, it was frustrating. His wage was still meager; nothing had happened in his career, and he was now going on twenty-seven. But Einstein did have a doctorate now and several more papers to his credit. He therefore decided to try for a privatdozent again. With his latest papers, he was in a much better position. He tried the University of Bern again, but worried that Aimé Forester wouldn't be impressed with his recent work.

Over the next few weeks he watched the mail. Finally the reply came. His application was rejected.

Chapter 11

Extending the Theory

Einstein settled back to work and put the University of Bern rejection behind him. But he soon had another problem to worry about. Walter Kaufmann of Göttingen had been attempting for several years to measure the change in mass of an electron that occurs when its velocity increases. This is what Einstein's theory predicted, but it was not the only theory that predicted a mass change. Three other theories gave similar predictions: Max Abraham of Göttingen and Alfred Bucherer had theories that gave different predictions, and there was as a theory by Hendrik Lorentz of Leiden that gave the same prediction as Einstein's, but it was not based on the relativity principle.[1]

Kaufmann had moved to Bonn by late 1905, when he made his announcement. According to his results, Abraham's theory was favored, and Bucherer's prediction was close, but Einstein's and Lorentz's theories were ruled out. Einstein's theory had, of course, done away with the concept of the ether, and Kaufmann went as far as suggesting that experiments on the ether should be resumed. To Lorentz, the announcement was a devastating blow, and he said that he was abandoning his theory. Einstein, however, was not so easily convinced. He wasn't impressed with Abraham's theory. It was

Fig. 17: Einstein at about thirty years old

not a universal theory in the way relativity was; a change in mass with increased relative velocity was only one prediction of many in his theory of relativity. In Abraham's theory it was the only prediction. Planck was also upset by the announcement that Einstein's theory was ruled out, but he advised caution before it was accepted. He said that more experiments were needed.

Although Kaufmann's results were close to Bucherer's, Bucherer was not satisfied. He set up his own experiment (which was a considerable improvement over Kaufmann's), and within a year he announced his results: Einstein's and Lorentz's predictions were extremely close; his and Abraham's were ruled out. He could not decide between Einstein's and Lorentz's theories since they made the same prediction (and it would be many years before it was possible to show which one was correct). He conveyed his results to Einstein, and Einstein was pleased. Einstein had, of course, been confident all along that his theory would be vindicated.

Einstein was still at the patent office in 1907, and he was becoming increasingly frustrated. His theory of relativity and other results were now getting considerable attention, but there was still no academic position in sight. He continued to worry that he might end up spending his life at the patent office (he was now referring to himself as a patent office "ink shitter"). A high school teaching job at Winterthur was advertised; he had taught there several years earlier, and he thought seriously about applying for it. Then a similar job became available at a Zurich high school, and he thought about applying for it. But in the end, he never applied for either job.

A VISITOR

Einstein was soon to receive a visitor. Max Planck's assistant in Berlin, Max von Laue, had become interested in Einstein's theory of relativity, having learned about it through a series of colloquiums that Planck set up. One day he went to Planck and told him he wanted to visit Einstein, assuming that he was at the University of Bern. He was surprised and slightly confused when Planck told him that Einstein was not at the University of Bern, but at the patent office in Bern.

Von Laue wrote to Einstein and made arrangements to meet him. He arrived at the patent office several days later. After asking the secretary to inform Einstein that he was there, he sat down in the waiting room and picked up a magazine. A few minutes later Einstein appeared at the door and looked around. Von Laue looked up from his magazine, but was sure the young, rather poorly dressed man couldn't be the discoverer of relativity; he was too common-looking. He returned to his magazine.[2]

Einstein went back to the secretary and told her that no one had asked for him. She checked and said he was still there. Einstein therefore approached the man with the magazine. "Albert Einstein," he said introducing himself. "Are you Max von Laue." Von Laue jumped up embarrassed and quickly apologized for not realizing who he was. They talked briefly, and Einstein invited him to come over to his apartment after he finished work.

As they walked toward the apartment, the two men talked about relativity. Einstein was pleased to hear that there was considerable interest in his theory in Berlin. As they continued on their way, Einstein offered von Laue a cigar. He took it and casually brought it up to his nose to smell it. He didn't like what he smelled and quickly put it in his pocket. Within a few minutes they were passing over the Aare River on a bridge. Von Laue reached in his pocket and took the cigar in his hand, then carefully, so that Einstein wouldn't see him, he dropped it in the river below.

Shortly after von Laue left, Einstein was surprised to get a reprint request from his old math teacher at the polytechnic, Hermann Minkowski. Minkowski had left the polytechnic in 1902 to go to Göttingen to work with renowned mathematician David Hilbert. The two men became interested in the foundations of physics, and after hearing about Einstein's paper on relativity, they began to study it. It's interesting that Minkowski told Max Born of Göttingen, "I really wouldn't have thought Einstein capable of this."[3]

Minkowski went to work on the new theory, trying to make it more mathematically elegant. He began by putting it in a four-dimensional formulation, assuming that in addition to the usual three dimensions of space, there was a fourth dimension—time.[4] He went on to show that Einstein's transformation was just a rotation and extension of the four space-time coordinates. Minkowski's formulation was beautiful and elegant, but it added nothing new to the theory itself. Many people found the new formulation delightful, but Einstein wasn't impressed with it. He didn't think it was needed and was sure it merely confused things. "Since the mathematicians have gotten hold of my theory I hardly recognize it," he said. But within a few years he had changed his mind and was using the new formulation himself.

The announcement of the new formulation was made first in Göttingen, then in 1908 at a conference at Cologne, which Einstein did not attend. Minkowski made his famous proclamation at the Cologne conference. "From now on space and time separately have vanished into the merest shadows, and only a sort of combination of the two now possess any reality."[5] We now refer to the union as space-time.

THE YEARBOOK ARTICLE

In September 1907 the editor of the "Yearbook of Radioactivity and Electronics," Johannes Stark, wrote to Einstein asking him to write an article on relativity for the following issue. Einstein had several ideas he was working on and felt that this was a good opportunity to get them into print. He had already realized that his theory of relativity was in conflict with Newton's theory of gravitation, and the reason, it seemed, was that his theory (Einstein's) applied only to straight-line, uniform motion. Einstein wanted to extend it to include accelerated motion, and in the process, he hoped it would also include gravity. The connection between accelerated motion and gravity came to him one day when he was sitting in the patent office; he later referred to it as "the happiest thought of his life."[6]

He was thinking about what would happen when a man jumped off a roof, when he realized the man would not feel his weight while in flight. He would, in effect, be weightless. This meant that there was a connection between acceleration and gravity; in short, the man would be accelerating, but he would have lost any feeling of gravity. Einstein was overwhelmed with happiness when he realized the connection; he later referred to it as the "equivalence principle." He knew that he might be able to formulate a new theory of gravitation with the new information, a theory that would replace Newton's theory of gravity. Newton's theory was based on the concept of a mysterious action-at-a-distance force that Einstein had had trouble accepting.

Einstein had also come to the realization that there was an equivalence between energy and mass. Earlier he had shown that there was mass associated with energy, but now he realized the relationship went both ways, and to every bit of mass there was an energy equivalent. This led to his famous formula $E = mc^2$ in which E is energy, m is mass, and c is the speed of light in a vacuum. Since the speed of light is exceedingly high (186,000 miles per second), its square is much higher, and the resulting energy from even a small amount of mass is therefore incredibly high (if indeed it could be converted to energy, and at that time there was no known way that it could).

Einstein applied his formula to a beam of light. If it had energy (and of course it was known that it did), it also had to have mass, and gravity would act on this mass. This told Einstein that a beam of light from a distant star would be deflected, or bent, if it passed near a gravitating body such as the Sun. But stars could be seen near the Sun only during an eclipse. Einstein determined the deflection and found that although it would be small, it should be measurable. He encouraged astronomers to check his prediction. A young astronomer in Berlin, Erwin Freundlich, became particularly interested in the experiment. He began by checking old plates to see if he could see any evidence of the effect, but he found that none of them were appropriate. Einstein would therefore have to wait for the next eclipse, and he was disappointed when he found that it wasn't until 1914.

A LETTER FROM KLEINER

Einstein was still trying to make up his mind about what to do about his future when he got a letter from Alfred Kleiner of the University of Zurich. For several years Kleiner had been trying to create a new position in the physics department to help lighten his load. He was now rector of the university and in a much better position to do something about it. Earlier he had encouraged Friedrich Adler, a former assistant, to apply for a privatdozent at the university, and Adler had completed the requirements. It seemed natural that he would assume the new position, a position that Kleiner now saw as a professorship in theoretical physics. Over the last year or two, however, Kleiner had become disillusioned with Adler. Adler's enthusiasm for physics appeared to be moderate, at best, and he was becoming increasingly interested in politics (his father was the founder of the Austrian Social Democratic Party). Furthermore, he was not a theoretician.

Kleiner began to think about Einstein. As his thesis director, he had been impressed with his thesis and was familiar with the work he had done since. He therefore wrote to Einstein encouraging him to apply for a privatdozent at the University of Bern, mentioning that if he did, a position at the University of Zurich might be available in the near future.

Einstein read the letter with interest. He had not heard from Kleiner in some time and was delighted to get a letter from him. The prospect of a professorship at the University of Zurich pleased him, but he had already applied for a privatdozent at the University of Bern and had been rejected. He was reluctant to tell Kleiner, but a few days later he wrote to him.[7]

Within a short time, he got a letter back from Kleiner telling him not to be discouraged. He told him to try again, but to go through the full habilitation this time, with its requirement of a thesis and a trial lecture.

He also mentioned that he would write a letter of recommendation for him. In just a couple of weeks, Einstein had written a thesis, collected the other materials he needed, and submitted them.

The situation at the University of Bern had changed since Einstein had last tried for a privatdozent. It had become increasingly evident to many of the officials that he was now one of the most famous physicists in Switzerland. In fact, he was one of the few who were actually known beyond its borders, and his output over the last few years had been impressive by any standards. Furthermore, it was also known that he had been turned down for a privatdozent twice, mostly because people in the physics department didn't understand his papers. This was becoming an embarrassment to them, and a number of them worried that it might be a poor reflection on the university. With the utmost speed, Einstein's application was processed. Shortly thereafter, everything was accepted, and in late February 1908 he became a privatdozent.

Einstein was pleased. He had looked forward to lecturing at the university, but he knew it wasn't going to help him much financially. He would receive no pay from the university, only a small fee from his students. Furthermore, he still had to work eight hours a day at the patent office, and he therefore had to squeeze his classes in at odd hours.

His first class was held during the summer session of 1908; it was offered at 7:00 in the morning. Only three students showed up, and all were personal friends: Michele Besso, Lucien Chavan, and Heinrich Schenk. Einstein was not comfortable lecturing, and with so few students, he didn't feel he had to do a lot of preparation, so everything was quite informal. The next semester he lectured in the evenings. His sister, Maja, was now attending the University of Bern and occasionally came to his lectures, even though they were far over her head.

Einstein soon began to find the lectures a hardship. He had to work all day at the patent office, prepare lectures as much as possible, and then deliver them, which was leaving little time for research. He began to wonder if it was worth it. To make things worse, occasionally he would arrive at the classroom to find only one student. But Kleiner had mentioned that a position might be available at the University of Zurich, and that kept him going.

To Einstein's delight, Jakob Laub arrived in Bern in March 1908. He had studied under Wilhelm Wien at Würzburg and had published two papers on relativity. Like von Laue, Laub had expected to find Einstein at the University of Bern, but when he arrived, Einstein was still at the patent office. "History is full of cruel jokes," he said.[8] The two men found so much to talk about that Laub stayed for three months, publishing two papers with Einstein. He became his first collaborator.

A PROFESSORSHIP AT LAST

Kleiner wanted to hire Einstein badly, but he was only one member of the committee that made the selection. In his proposal he stated that Einstein was "one of the most important theoretical physicists of the day."[9] He also discussed his impressive publishing record. But there was a problem: he was Jewish. To counter this, Kleiner pointed out that "all Jews were not the same," and in his experience with Einstein, he had "detected no undesirable personality traits."

Kleiner also had to give a recommendation in regard to Einstein's teaching, so he had to attend one of his lectures. He therefore traveled to Bern to observe him in action. Einstein was not prepared for the visit and had only one of two students in attendance when Kleiner came into the classroom. Knowing that he was on trial made him very uncomfortable; furthermore, with so few students, he had not prepared properly. Kleiner went away disappointed. He was not impressed with Einstein's teaching ability and mentioned it to several people at Zurich. Among them was his assistant Friedrich Adler, and through Adler, the remarks got back to Einstein. Einstein wasn't happy about it, so he wrote to Kleiner. At this stage, incidentally, Adler had withdrawn his name from consideration. "If it is possible to get a man like Einstein for the university, it would be absurd to appoint me," he wrote to the committee. "I must frankly say that my ability as a researcher does not bear even the slightest comparison to Einstein's."[10]

Despite Einstein's poor performance in the classroom, Kleiner still wanted to hire him. He was sure that his teaching would eventually improve. He therefore told Einstein outright that if he could improve his teaching, he would give him a very strong recommendation. But he would need another trial. Kleiner suggested that he give a talk at the Physical Society of Zurich in mid-February 1909, and Einstein agreed. Einstein gave the lecture and did a fairly good job. "I was really lucky. Totally against my usual habit I lectured well," he said after the talk.[11] Kleiner agreed that the lecture was an improvement over his earlier one and gave him a strong recommendation, as promised. When the vote was finally taken it was ten for Einstein and one abstention. Within a short time, Einstein had a letter from Kleiner offering him a professorship at the University of Zurich. But when Einstein saw the wages that they were offering (they were considerably lower than what he was making at the patent office), he declined it. Kleiner therefore went back to the central committee, and they soon agreed to match Einstein's current wage of 4,500 francs per year.

Einstein resigned from the patent office on July 6, 1909, effective October 15. When he told Haller he was resigning, Haller was surprised and thought he was joking. He couldn't believe that Einstein had been offered a professorship at the University of Zurich.

THE FIASCO

The announcement that Einstein had accepted a position at the University of Zurich was placed in several newspapers. Anna Schmid, the young girl Einstein had written a poem to at the Hotel Paradise many years earlier, saw it and wrote to Einstein congratulating him. She was now married to George Meyer and living in Basel. Einstein was delighted to hear from her. He wrote back, "I probably cherish the many lovely weeks that I was allowed to spend near you in Paradise even more than you do."[12] He also invited her to come to see him in Zurich and gave her the address of his office at the university. Anna wrote back immediately, but somehow the letter was intercepted by Mileva. Suspecting that Einstein was having an affair, she took action. She wrote to Anna's husband, telling him about the affair and how outraged she was. She said that Einstein himself was also outraged that Anna had written to him and was carrying on as she was.

When Einstein heard about this, he almost went through the roof. He wrote to Meyer immediately and apologized, saying that nothing had ever occurred between him and Anna. Indeed, they had not seen one another in years. Einstein was so beside himself about Mileva's behavior he barely spoke to her for weeks, and it put even more pressure on their already strained relationship. Months later he wrote to Besso that his "spiritual equilibrium" had still not been restored.

THE HONORARY DOCTORATE

Einstein was still at the patent office when he got a large envelope in the mail. Opening it, he found the letter inside was in colorful script and was mostly in Latin. Thinking it was some sort of advertisement, he threw it in the wastebasket. As it turned out, it was an invitation to help celebrate the 350th anniversary of the birth of the University of Geneva. Along with Marie Curie, Wilhelm Ostwald, Ernest Solvay, and many other dignitaries and scientists, he was to receive an honorary doctorate.[13]

The officials at the university waited for a reply from Einstein, and when they didn't get one, they began to worry. They got in touch with Einstein's friend and former student Louis Chavan, who was from Geneva, and asked him to make sure Einstein came to the celebration. Chavan convinced Einstein to go, but he didn't tell him why. And when Einstein found out he was to receive an honorary doctorate, he was shocked. He didn't have a proper gown or cap, nevertheless he got in the processional line with everyone else, wearing a suit and a straw hat. Although he looked out-of-place, it didn't bother him, and he thoroughly enjoyed himself while he was in Geneva. A gigantic dinner was served at one of the large halls after the ceremony, with Einstein saying that he felt guilty about eating so much. This was the first of many honorary degrees he would get over the years.

An ironic twist of the affair is that one of the other persons getting an honorary degree was Wilhelm Ostwald—the same professor Einstein had written to asking for an assistantship only a few years earlier. Indeed, Einstein's father had also written to him, and neither had received a reply.

THE NEW PROFESSOR

The Einstein's moved to Zurich in mid-October 1909 and found an apartment at 12 Moussonstrasse. Einstein discovered to his delight that Friedrich Adler was in the same building, on the floor below him. The two men soon renewed their friendship. As it turned out, Adler's relationship with Einstein would help him in years to come. Adler followed in his father's footsteps and became very politically active. In fact, it's safe to say that he became somewhat of a radical. Two years after World War I began, he walked up to the Austrian premier, Karl Stürgkh, and after blaming him for prolonging the war, he shot him to death. He was brought to trial and condemned to die. Einstein wrote a passionate letter on his behalf that helped get his sentence commuted to the rather light sentence of only eighteen months.

As Einstein assumed his professorial duties, he continued to worry about his teaching abilities, but he now spent a lot of time preparing his lectures. At first, his students weren't quite sure what to make of him. His suit was unpressed and ill-fitting, and he seemed nervous. But it didn't take long before things changed. Einstein soon began to relax before his students, and his teaching technique improved. He began to enjoy his students, and they enjoyed him in return. They soon found that he was quite different from most professors, who liked to keep their distance from the students. He was friendly and encouraged questions and discussion in class. This was something new to them. In fact, according to one of his students, he was usually surrounded by students after class and frequently took them to one of the cafés in Zurich where they continued discussing science. There was no doubt that Einstein loved to talk about science. On a few occasions, he even took students back to his apartment with him.

His relationship with Mileva hadn't improved much, and with his students he had an excuse for avoiding her. At this point, however, things had not yet gotten out of hand. They were still speaking to one another. In fact, it was shortly after Einstein began at the university that Mileva found out she was pregnant again.

With his heavy teaching load, Einstein had little time for research, but he squeezed in as much as possible. He was still working on his extension of relativity theory, but making little progress. His prediction of the bending of a light beam from a star as it passed near the Sun had not yet been tested. He was also working on quantum theory. Although he found it difficult to get much research done, he now had an assistant by the name of Ludwig Hopf. An added benefit was that Hopf was an accomplished pianist and frequently accompanied him in duets. The two men published two papers together.

Many of the people who knew Einstein at this time noticed a significant change in his personality after he became a professor at Zurich. He had always been confident of his abilities, but with so many rejections, he had had a chip on his shoulder. He was what many would describe as "an angry young man," with a disdain for the scientific establishment. He felt like an outsider who was continually having to prove himself. It's perhaps a little ironic that he was now becoming the "establishment" that he had earlier rebelled against. His opinions were now valued; he was asked to present papers, and well-known scientists from throughout Europe were eager to meet him. As a result, the mistrust and uncertainty in his personality disappeared, and he became a much more confident person. Despite his growing fame, however, he remained humble and friendly, and he got along well with the other faculty at the university. Furthermore, his students loved him.

Einstein was only at the University of Zurich for about six months when he began to get inquiries from the German University at Prague. A full professorship would be opening, and they wondered if he was interested. The wages would be double what he was presently getting, and his teaching load would be much lighter. Einstein was uncertain about Prague, but he was interested. The increased wages and increased time for research were an enticement. But it soon became clear that there were problems.

As it turned out, the inquiry from Prague worked in his favor. His students found out that he might be leaving, and they signed a petition asking the department of education to do anything they could to keep him in Zurich. And it helped. Almost immediately he got a raise of 1,000 francs a year, increasing his salary to 5,500 francs a year, and there was a directive to lighten his teaching load. Then, for several months Einstein never heard from Prague.

As Einstein spent more and more time with his students, his teaching and research, Mileva fell into a greater depression. She was sure that Einstein was ignoring her and wrote of her feelings to Helene. "You see I am starved for love and would be so overjoyed at having a yes that I almost believe wicked science is guilty," she wrote.[14] She became increasingly close to her son, Hans. "We are actually inseparable and cling terribly to each other," she said.[15] Adding to her misery were occasional visits from Einstein's mother. The atmosphere was extremely tense when she was in the house, and as a result, her visits were rare. Hans stated later in life that he barely knew his grandmother. He also said of his mother, "She was a typical Slav, capable of strong negative feeling and once hurt she could not forget."[16]

During her depression Mileva gave birth to Eduard. Born on July 28, 1910, he was soon nicknamed "Tete." The birth helped for a while, but it wasn't long before things were back to where they had been. Mileva was particularly jealous over Einstein's relations with women, and indeed by this time she had reason to worry. Einstein's assistant, Ludwig Hopf, recounts an occasion when he, Einstein, and several other people went to a café after a talk they had all attended. In the group were two extremely attractive sisters. Einstein sat most of the evening ogling them, hardly joining in the conversation, which seemed to bore him. Hopf finally scolded him, and Einstein, looking a little sheepish, apologized.

Eventually the problems at Prague began to settle. Einstein's main competition for the job was Gustav Jaumann, a professor at the Technical Institute at Brno. He was Austrian and Einstein was a Jew, so he had considerable support from the anti-Semitic officials at the university. But when they ranked the candidates according to their accomplishments,

Einstein was placed at the top, with Jaumann second. When Jaumann heard this, he became outraged and demanded an exorbitant wage, if he were to be hired. He was soon dismissed from consideration, and Einstein was offered the job. But there was one final hurdle. In the application form under religion Einstein had written the word "none." Officials at Prague would not accept this, so Einstein changed it to Jewish, and this seemed to satisfy them.

Einstein was offered the job in January 1911. He would be a full professor at the young age of thirty-one, and his wage would be almost double what he had received at the University of Zurich. Furthermore, he would have more time for research. Einstein was pleased and looked forward to going to Prague. Mileva, on the other hand, was apprehensive and worried.

TO PRAGUE

Mileva's mother came to help them move, and the family arrived in Prague on April 1, 1911.[17] Mileva detested the city almost from the beginning. Einstein liked it at first, but as he became more familiar with the people, he found many things he didn't like. The city was, indeed, a strange mixture of cultures. About 90 percent of the population was Czech, with the German overlords making up the remaining 10 percent. For the most part, the Germans looked down on the Czechs, so there was considerable friction between the two nationalities. To further confuse things, there was a large community of Jews among the Germans, and they kept to themselves.

The Czechs and the Germans rarely mixed with one another, so Einstein had little contact with the Czechs personally, but he did encounter them on the streets, and he found them sullen, with a strong resentment for the Germans. Aside from the conflict between the two groups, the water and air were inferior to that of Switzerland. The water was brown, stained the sinks, and was generally undrinkable. Einstein had to get drinking water from the public fountain. Furthermore, there was considerable ash and soot in the air, and there appeared to be many more bugs, fleas, and so on than there were in Switzerland. And finally, it was much hotter than they were used to.

For the most part, however, Einstein liked the university. He was treated as a celebrity, and his inaugural speech was held in the largest auditorium on the campus. It was attended by everyone of importance at the university. He also had a large office and had brought his assistant, Ludwig Hopf, with him. In addition, the large library impressed him. But within a short time, he began complaining about the students. He found them much less industrious and conscientious than his Swiss students.

Einstein had more time for research, but aside from Hopf, he had no one to talk to, and after a couple of months, Hopf left. The major problem, however, was that he was now facing a serious problem. He soon found that in his quest for an extension of his relativity theory, he was going to have to abandon Euclidean geometry. His friend George Pick of the mathematics department suggested that a new branch of mathematics called tensor analysis might be helpful, but there is no evidence that Einstein looked into the suggestion at this time.

Einstein was also working on quantum theory. And his feelings for it were evident. Several times when he had visitors in his office, he would take them to the large window at the end of his office. It overlooked the beautiful gardens of an insane asylum next door, and the inmates would frequently walk around in it. Einstein would say to his visitor, "Here you see the crazy people who are not working on quantum theory." (The "other" crazy people are working on it!)[18]

Despite the hardships in Prague, the Einsteins' standard of living increased in several ways. They now had electric lights, and Einstein's wages were high enough that they could hire a live-in maid. As it turned out, the maid discovered shortly after she began work that she was pregnant, and the baby's father was long gone. This had to have reminded Mileva and Einstein of their predicament several years earlier, and they were quite sympathetic. They allowed the maid to have the baby in their apartment, and they also helped when she put it up for adoption, and finally when this didn't work, they allowed her to keep it with her.

During the fall of 1911 Einstein attended the first Solvay conference in Brussels. It was organized and sponsored by Ernest Solvay, a rich chemical industrialist. The idea was to bring the best scientists of Europe together so that they could share their ideas and keep each other abreast of the latest developments. The conference was given considerable press coverage, but to the dismay of Sovay, the press concentrated on something that created a sensation. Marie Curie was among the attendees; she was now a widow, as her husband, Pierre, had been killed when he was run over by a carriage in 1907. Also present was Paul Langevin of Paris. For several years Langevin had been having trouble with his marriage and was in the process of trying to get a divorce. In the meantime he and Marie Curie had been having an affair in Paris. Langevin's wife had discovered some of their letters, and she gave them to the press just before the conference. Soon they were front-page news; the press accused Curie of trying to entice Langevin away from his wife and four children, and the story wouldn't go away.

Asked for his opinion of the affair, Einstein said he didn't believe it. He had seen them together several times, and there didn't appear to be anything going on between them. Commenting, he said, "Madam Curie . . . is a simple, honest woman, almost buried under her duties and obligations. She has sparkling intelligence but, despite her passionate nature, is not attractive enough to be a danger to anyone."[19]

During the summer of 1911 Einstein returned briefly to Zurich to teach a short course. Marcel Grossmann was now at the polytechnic; indeed he was dean, and after meeting with Einstein, he asked him if he was interested in a position at the polytechnic. Also, unknown to Einstein, Heinrich Zangger, the dean of forensic medicine at the University of Zurich, pressed the polytechnic administration to hire Einstein, and actually went over the president's head when he expressed some reluctance. To Einstein, this had to be a little ironic. This was the polytechnic where he had been rejected by several professors for assistantships only a few years earlier; now they wanted him as a full professor. Einstein, of course, was interested, and negotiations proceeded after he returned to Prague. When Mileva heard of the possibility, she was delighted and eager to return to Zurich. As usual, however, there were snags in the negotiations, but in February 1912 an offer was made and Einstein accepted it.

Before Einstein moved back to Zurich, he traveled to Berlin, where he visited his cousin Elsa, whom he had not seen since childhood. She was the daughter of Rudolf and Fanny Einstein. Divorced, she had two daughters and lived in the same apartment building as her parents. Einstein's mother had lived with Rudolf and Fanny for a while after Hermann's death, but she was no longer there. Einstein began corresponding with Elsa after he returned to Prague, and also later at Zurich, and this time Mileva did, indeed, have something to worry about.

Chapter 12

The General Theory

While still in Prague, Einstein realized that significant changes would be needed if his special theory of relativity was to apply to all types of motion. Furthermore, as a result of his principle of equivalence, he knew that if his theory was generalized to include accelerated motion, it would also be a theory of gravity.

By 1912 Einstein had adopted Hermann Minkowski's four-dimensional formulation of his theory, and through a study of the rotating rigid disk, had come to the realization that the usual geometry of three-dimensional space, namely, Euclidean geometry, would not work in his generalization. As is well known, any disk has a fixed ratio between its diameter and its circumference; it is known as π and is 3.14159. According to special relativity, however, an object shrinks in the direction of its motion as relative speed increases, but perpendicular to its motion there is no shrinkage. This means that in the case of the rotating disk, each little section along its circumference will shrink as it rotates faster and faster. Its diameter, however, will remain the same, and therefore the ratio between the diameter and the circumference cannot be π.[1] Einstein's friend, Paul Ehrenfest, published a paper on this paradox shortly after Einstein discovered his special theory of relativity, and it became known as the Ehrenfest paradox.

Einstein spent a lot of time in Prague thinking about this paradox, and he soon realized that an extension of his theory would require a change in the geometry of space-time. But how would the new geometry differ? He had studied Carl Gauss's theory of surfaces under Carl Geiser at the polytechnic, and it seemed that it was a step in the right direction, but it was only a two-dimensional theory. He would need four dimensions.

At this stage Einstein was still unsure of himself in relation to his new theory. He had many pieces of the puzzle. For example, he had established the principle of equivalence, knew that the theory should go over to Newton's theory (give the same predictions) in the limit of weak fields, and was convinced that a beam of light from a star would be affected by the Sun—in effect, it would bend around the Sun as it passed near its limb. Furthermore, he was sure he would be able to explain an anomaly in Mercury's orbit with his new theory. Newton's theory did not explain Mercury's motion exactly, and a number of astronomers believed that it was being perturbed by a small planet inside its orbit called Vulcan.[2]

What Einstein desired was an all-encompassing theory, namely a set of equations that would describe the gravitational field and how a massive object moved in it, and at the same time, would solve all of the above problems. In addition, he wanted the equations to be covariant, in other words, independent of the coordinate system. In this case all predictions would be the same regardless of what coordinate system was used to set up the problem.

Through the efforts of Zangger and Grossmann, Einstein was now on his way back to Zurich as a full professor. He was going to the polytechnic, his old alma mater. In April, before he left Prague, he visited Berlin where he was reunited with his cousin Elsa. He had not seen her for many years. She had been divorced a few years earlier and had two daughters, Ilse and Margot. Einstein was enamored with her almost immediately and spent a couple of days with her at the Wansee resort area, near Berlin. We don't know what he said to her while he was in Berlin, but he couldn't wait to write to her when he got back to Prague. And naturally this time he was more careful; the incident with Anna Schmid had made him cautious, and he made sure Elsa sent her letters to his office at the university. He was excited about the affair, and it wasn't long before he declared his love. But he must have felt guilty because he was soon having second thoughts. Though he knew his marriage to Mileva was on the rocks, nevertheless, he was still married and had two children. Within a week, he wrote, "There would only be confusion and misfortune if we were to give in to our mutual attraction."[3] Two weeks later he wrote to her informing her that he would no longer be writing to her, and he urged her not to write to him. "I have the feeling that nothing good will come of it, not for the two of us, and not for others if we were to get close to each other."[4]

PROFESSOR AT THE POLY

The Einstein's moved to Zurich on July 15, 1912, and soon found an apartment at 116 Hofstrasse. It was much larger than the apartment they had lived in when they were in Zurich earlier. After Prague, Einstein was glad to be back in Switzerland, and Mileva, it's safe to say, was even more delighted than he, and relieved. She had hated Prague, and Zurich was like home to her. Einstein was also happy because both Marcel Grossmann and Louis Kollros from his student days were now at the polytechnic. Furthermore, his friend Max von Laue had taken a position at the University of Zurich, which was nearby.

Soon after Einstein was settled he called on Grossmann. "Grossmann, you've got to help me," he said with despair in his voice.[5] He was desperate because he knew that in his new theory he would have to deal with non-Euclidean geometry, and he wasn't quite sure how to go about it. He remembered, however, that Grossmann had done his thesis on non-Euclidean geometry, and hoped he would he able to help. Einstein explained the problem to Grossmann and was delighted when he seemed interested. He had one stipulation, however; he would help only with the mathematics. He wanted nothing to do with the physics, which would be up to Einstein. Einstein, of course, quickly agreed.

The first problem was a generalization of Gauss's theory of surfaces to higher dimensions. Grossmann's familiarity with the area was limited, but over the next few days, he looked into the problem and found that considerable work had been done. The search took him to a branch of mathematics called tensor analysis, and he warned Einstein that it was exceedingly difficult.[6] They would be dealing with tensors (mathematical concepts that change in a specific way when the coordinates are changed), which have many components. The equations of the gravitational field would, in fact, be written in terms of tensors. But considerable work had been done in the area by Bernhard Riemann, Curbastro Ricci, and Tulio Levi-Cevita. Indeed, Riemann had formulated a tensor that could describe any surface in any number of dimensions. Einstein was mainly interested in the curvature of a "surface" of space-time, but the Riemann tensor would still work. Furthermore, it was local; in other words, it could describe the curvature even if it varied from point to point.

Einstein was overjoyed. He knew they were on the right track. His passion for the new theory was at its peak, and he was certain a breakthrough was imminent. Einstein wanted his new equations to be covariant, and Grossmann soon showed that an offshoot of Riemann's tensor, namely Ricci's tensor, was more appropriate in this case. With a little trial and error, they soon arrived at an equation for the gravitational field. All they had to do now was check that in the limit of weak gravitational fields that it went over to, or gave the same results as, Newton's gravitational equations. When they checked, they were disappointed to find that it didn't. The equation was simple and covariant, but it didn't appear to give them what they wanted. Einstein was frustrated, but reluctant to give it up. At last, however, he did, and when he looked for a similar covariant equation, he couldn't find any.

The two men then took the only alternative that seemed available to them. They decided that the theory could not be covariant. This would mean that the equation would depend on the frame of reference, and this bothered Einstein. Nevertheless, he finally resigned himself to it and spent considerable time trying to prove that the theory had to be noncovariant. He came up with two different proofs, and this seemed to satisfy him. Deep down, though, he wasn't happy. With this settled, he and Grossmann came up with a noncovariant theory and published it under the title "Draft of a Generalized Theory of Relativity and a Theory of Gravity." The paper was divided into two parts, one labeled the physical part and one the mathematical part. This was for Grossmann's benefit, for as expected, Einstein was the author of the physical part and Grossmann the author of the mathematical part.[7]

As we will see later, there was considerable irony in discarding the first (covariant) equation. Einstein had made a mistake in checking it to see if it gave the same results as Newton's theory. It was, indeed, identical to Newton's theory in the limit outside matter, but it took two frustrating years to find this out.

Einstein worked hard while he was in Zurich, but his life wasn't all work. He played musical duets with his old math teacher Adolf Hurwitz. Interestingly, this is the same Dr. Hurwitz who had turned down his application for an assistantship several years earlier. The Einsteins spent many "musical evenings" at the Hurwitzs' and their daughter Lizbeth became a good friend of Mileva's. Hans was now playing the piano and occasionally joined in the music-making. Einstein was always jovial and lighthearted during these sessions, but Lizbeth Hurwitz later said that Mileva became increasingly gloomy during them and sometimes sat by herself all evening saying almost nothing. Knowing she particularly liked Schumann, they would occasionally arrange to have a "Schumann evening."[8]

Einstein turned thirty-four on March 14, 1913, and to his surprise he received a card from Elsa.[9] Life with Mileva was becoming increasingly burdensome to him, and despite the pledge he had made many months earlier, he was delighted to hear from her. She wished him a happy birthday, asked if there was a layman-level book on his theory that she could read, and asked him for a photo. Einstein was ripe for the plucking. He wrote back eagerly, explaining that there was no layman book on relativity, but he would gladly explain the theory to her sometime. He even went as far as inviting her to come to Zurich. So much for his pledge to stay away from her.

By this time Einstein's marriage had hit a new low, and he and Mileva were barely talking to one another. Hans later said that he noticed that there was a difference in their relationship after 1912. What Einstein wanted at this point, more than anything, was sympathy; in effect, he wanted a shoulder to cry on, and in Elsa he certainly got it. Not only could he tell her his problems with Mileva, but he could tell her his problems with his mother. Elsa had lived close to Einstein's mother for several years and had not gotten along well with her, so she was quite sympathetic when he complained about her, too.

On March 27 Einstein and Mileva traveled to Paris where Einstein visited with Solovine, Langevin, and Marie Curie. They stayed at the Curies' and became close friends. Marie Curie had two daughters, one about the same age as Hans. They enjoyed their stay at the Curies' so much that they made arrangements to meet again during the summer, when they would go on an extended hiking tour through the Swiss Alps.

VISITORS FROM BERLIN

Einstein had only been in Zurich for a couple of semesters when the gears in Berlin began to grind. Fritz Haber, the director of the newly formed Kaiser Wilhelm Institute of Physical and Electrochemistry, wanted Einstein in Berlin. It was now well known that Einstein was the brightest of the up-and-coming young scientists in Europe, and Haber convinced the Prussian Academy that it was in their best interest to make him an offer that he couldn't refuse. Incidentally, this is the same Haber who later became notorious as the developer of mustard gas and other deadly gases that were used in World War I, and in World War II for gassing Jews. The irony of this is that he was a Jew; strangely, though, he did everything possible to disguise the fact.

Max Planck and Walther Nernst were sent to Zurich as emissaries. They arrived in late July. It might seem strange that the Germans were so interested in Einstein. After all, he was Jewish, and Jews were generally looked down upon by aristocratic Germans; furthermore, he had at one time renounced his German citizenship (it's unlikely, though, that they looked into that). It's also unlikely that they knew he had a deep distrust of Germany. Planck was fully aware of Einstein's abilities and knew what his potential was, but he was quite wary of his attempts to generalize special relativity and create a theory of gravity. "First of all it won't likely work, and if it does no one will believe you," he told Einstein.[10] Planck also had reservations about Einstein's idea of "quantum radiation," even though he had originated the idea of the quantum. Addressing the Prussian Academy, he said that Einstein "sometimes goes off the deep end [he was referring to quantum radiation] but that shouldn't be held against him."[11] Despite the reservation, however, Planck was as anxious as anyone to lure Einstein to Berlin.

Einstein heard rumors and knew Berlin was interested in him, but what amazed him was the generosity of the offer. As in Zurich, he would be a full professor at the University of Berlin, and his wage would be 12,900 marks, which was about one-third greater than his wage at Zurich. Indeed, it was the maximum allowed for any professor at the university. He would also be the director of the Kaiser Wilhelm Institute of Theoretical Physics when it was built and best of all: he would have to do no lecturing and could devote himself totally to research.

Einstein was overwhelmed. He hated to leave his beloved Switzerland, and he knew that Mileva would dread a move to Berlin. But the offer was tempting. The increase in wage wasn't important to him, since his wage was already high and he had a certain disdain for money. Furthermore, it's unlikely that the directorship of the institute appealed to him. But what did appeal to him was the free time he would have to work on his research. He had a relatively heavy teaching load at the polytechnic, and it was beginning to wear him down. His major interest in life, at this point, was his new theory, and if he would have more time to work on it, he would have to take the offer seriously. The idea of returning to Germany didn't appeal to him; he had left Munich many years earlier and vowed never to return. But the University of Berlin was a powerhouse in physics; physicists such as Planck, Nernst, Wien, and Sommerfeld would be nearby.

Planck and Nernst gave Einstein a day to think over the offer. They went sightseeing by train and told Einstein that they would see him the following day. With his usual humor, Einstein told them that if the answer was yes, he would wear a red rose in his lapel when he met the train.[12] If it was no, he would wear a white rose. As expected, Mileva was crestfallen when Einstein told her the news. Several of Einstein relatives lived in Berlin, and there was Elsa, of whom she was already becoming suspicious. In addition, she didn't know anyone in Berlin, and Slavs were generally looked down on by Germans. But her arguments fell on deaf ears, and when Einstein met the train the next day, he had a red rose in his lapel.

Einstein could have left the following fall, but he decided to delay until spring. The Einsteins had agreed to go hiking with the Curies during August, but just before they were to leave, Eduard got sick and Mileva had to remain behind with him. Einstein, Hans, Marie Curie, and her two daughters, Eve and Irène, went on the hike. Einstein spent a lot of time talking to Marie, no doubt about his new theory, but as an experimentalist, she was in no position to give him advice. Nevertheless, the hike was a success. Mileva and Eduard later joined the group, but Mileva did not hike.

Einstein wrote to Elsa about the trip as soon as he got back to Zurich. Marie Curie was twelve years older than Einstein, but Einstein must have worried that Elsa would consider her a romantic threat. He praised her intelligence and wit, but said that she complained continuously. "She has the soul of a herring," he wrote to Elsa.[13]

For several years Einstein had been anxious to have his new theory tested, by measuring the bending of a beam of starlight near the rim of the Sun. He was delighted when Erwin Freundlich of Berlin offered to help. Freundlich had, in fact, now offered to lead an expedition to Siberia for the next eclipse, which would occur in July 1914. Freundlich's presence in Berlin was another reason that Einstein accepted the Berlin offer.

In September Mileva managed to convince Einstein to go on a holiday. She wanted her family to see Eduard; they had seen Hans earlier, but had never seen Eduard. Reluctantly, Einstein accompanied Mileva to her family's farm near Novi Sad. During the trip the two boys were baptized in the Christian Orthodox Church. Einstein was indifferent to the religious rite and didn't attend it; furthermore, he and Mileva returned to Zurich via different routes.

Einstein went to Berlin to talk to Planck, Haber, and others about his new position, but his main reason for going, no doubt, was to see Elsa.

Elsa overwhelmed him with her cooking and mothering, and she lent a sympathetic ear to his complaints about Mileva. He complained about the "icy silence" between them. Elsa gave him a hairbrush and a toothbrush, teasing him that he should improve his appearance. He used the hairbrush but refused to use the toothbrush, saying that the "boar bristles in it could drill through diamond."[14] Einstein took this teasing good-naturedly, but there are indications he was wary of too much mothering, even then.

In October Einstein attended the Solvay conference in Brussels, but he begged off presenting a paper, saying he was too busy. He may have been disappointed that there was little interest in his new theory. Most of the interest was directed at the new quantum theory that Niels Bohr of Denmark had just presented. Einstein praised the theory, but as we will see, he later had many doubts about quantum theory.

As the time to leave for Berlin approached, Mileva became increasingly gloomy. According to Einstein, she complained continuously about the impending move. One of the reasons for her worry was that she would likely see more of Einstein's mother. Pauline had still not accepted Mileva as her daughter-in-law, and she and Mileva had been tiptoeing around one another for years. Mileva dreaded every visit she made. Making things worse was that Pauline was usually quite tactless. The split finally came near Christmas of 1913. Pauline sent a letter addressed to Einstein alone, telling him that Christmas presents would soon arrive. They were not sent directly, however, but were sent to Maja to deliver, and Mileva was left in the dark. Mileva was outraged that she had not been told, and returned the presents immediately, informing Pauline that the boys would no longer be seeing her. And indeed it was probably the last visit between them. Einstein didn't intervene; at this stage his attitude was "they deserve each other."[15]

Einstein didn't worry about his accommodations in Berlin, but he knew that Mileva would be fussy about where they lived. He therefore sent her to Berlin on December 28 to look for an apartment. She stayed with the Habers, and they helped her look for an apartment. Fritz's wife, Clara, was particularly helpful and became fond of Mileva. Elsa heard she was in Berlin and offered to help, but Mileva quickly refused her offer.

As the date for the move approached, Einstein was excited, but he was a little apprehensive. "They are betting on me as a prized hen," he said. "But I'm not sure I can lay any more eggs."[16]

MOVING TO BERLIN

Einstein arrived in Berlin on April 1, 1914. Eduard got sick just before the move, and his doctor advised that he needed a rest. Mileva took him to Locarno, so she and the boys didn't arrived until the end of April. Einstein took advantage of his weeks alone to visit frequently with Elsa, and even after Mileva arrived, he would disappear for long stretches of time, with no explanation. It was soon obvious to Mileva who he was visiting, and the tension between them became even greater. Hans, who was ten at the time, said that his parents hardly spoke to one another, and he could feel the tension between them. Einstein's feeling of frustration about his marriage came out in his letters to Elsa. "She's an unfriendly, humorless creature who herself has nothing from life and who undermines others joy of living through her mere presence," he wrote.[17] On another occasion: "She's the sourest sourpot there has ever been."[18] But Einstein admitted that he knew the real reason for her gloominess. She was never "good with people," and aside from Helene, had few friends she could confide in. She clung to Einstein (perhaps too much), and when he lost interest in her, she despaired and went into depression.

In May Paul Ehrenfest visited the Einsteins in Berlin and found Mileva very gloomy. It was obvious that her fears had come to pass; Berlin was even worse than she had anticipated. Einstein, on the other hand, accommodated easily. "Things are nice here in Berlin," he wrote to Ehrenfest shortly after getting settled. He had a large office at the Prussian Academy and also spent considerable time at the well-equipped library. Hans was now in school, and as Einstein had many years earlier, he hated the discipline of the German school. Mileva couldn't wait for the school term to end, but she didn't have to wait. She and Einstein had a serious argument, with Mileva accusing him of being a "patsy" to the Germans. She had also confided her troubles in Clara Haber, which irritated Einstein. In a rage, he moved out to his uncle's apartment.

Mileva went to Clara Haber with her troubles again, and Clara invited her and the boys to move in with them temporarily. Einstein was furious. Over the next few days, Fritz Haber (Einstein's boss!) carried notes back and forth between Einstein and Mileva. Einstein wrote down a list of conditions that Mileva would have to abide by if they were to live together. At first she agreed to them, then changed her mind. Finally, on July 24 a contract was drawn up for their separation. Einstein would pay her 5,600 marks (7,000 Swiss francs) support for the children. She would keep the boys and move to Zurich; Einstein could visit them, but only on "neutral ground," and never at Elsa's.[19]

SEPARATION

Michele Besso came to take Mileva and the boys back to Zurich. Fritz Haber went along with Einstein to the train station to see the boys off. Mileva apparently made a last effort for a reconciliation at the station, but Einstein wasn't interested. He had had enough and knew it would never work. He had strong feelings for his boys and was heartbroken when they left. Haber had to help him from the station, and he was in tears for the rest of the day.

The Habers must have wondered what they had got themselves into. Fritz had been determined to get Einstein to Berlin, but it's unlikely he knew about his marital problems. It was also difficult with Mileva confiding in Clara, and Einstein telling his troubles to her husband. Besso, of course, had known about the problem for years and was also in a difficult position. He was very sympathetic to Mileva, but had been friends with Einstein for twenty years. Furthermore, even when he tried to help, Mileva was very cautious of him because of his friendship with Einstein.

Mileva found temporary quarters in a boarding house when she got to Zurich, and from the beginning, she complained to Einstein that the money he was giving her was not enough. When he was at the patent office, he was making only 4,800 francs a year, and now she was getting 7,000. At one point she read that Einstein had received a prize of 1,000 francs for some work that he had done, and she to wrote him demanding that he send some of it to her. Einstein flew off the handle and told her that he was sending her enough money (he was sending her approximately half his rather high wage, and at the same time he was also sending his mother money). Over the next few years, he did, indeed, frequently send her additional money. Still, it wasn't enough, so to make ends meet, Mileva began tutoring in mathematics and teaching piano.

As Einstein's personal life churned in turmoil, Europe itself was becoming a cauldron of hatred and strife. The Austrian Archduke Francis Ferdinand was assassinated by a Serbian on June 28, and within a short time, Germany had declared war on Serbia. Russia, as a backer of Serbia, was soon involved, then France and England joined. Then on August 4 Germany invaded Belgium, and rumors of atrocities circulated around the world.

Einstein, a strong pacifist, was dumbfounded. He was particularly upset because Freundlich and the astronomers in his group, who had journeyed to observe an eclipse in Siberia to check on his prediction of the bending of light, were taken prisoner. Fortunately, within a short time, they were traded for some Russian officers who had been captured, so Freundlich was back in Berlin by Christmas, but all his equipment had been confiscated. The only consolation was that it was so cloudy the day of the eclipse that the expedition would not have gotten any photographs anyway.

With world opinion strongly against Germany, particularly after it invaded Belgium, a number of prominent Germans drew up a manifesto claiming that Germany had no alternative, and its motives were justified. Furthermore, according to it, no atrocities had occurred in Belgium. Ninety-three leading scholars, scientists, and politicians signed the manifesto. Among them were Planck, Nernst, and Haber. Since Einstein was still a Swiss citizen, he wasn't asked to sign it, and if he had, he certainly would have refused. He was disappointed that so many of his colleagues had signed it. Planck, however, later had reservations about it; one of his nephews was soon killed, and his son was taken prisoner. Furthermore, both of Nernst's sons were killed in action.

In an effort to provide some balance, a friend of Elsa's, Georg Nicolai, circulated a countermanifesto, urging a stop to all the nonsense. There would be no victor, only victims, he said. Einstein agreed and quickly signed it, but few others did. Nicolai had difficulty getting it published, so it was never widely circulated.

Despite the frustration of war, Einstein wrote to his friends that he was enjoying the single life. He now had considerable free time to work on his theory, and he was free from the frustrations of an unpleasant marriage. He had moved to a smaller bachelor apartment, a short distance from Elsa's. It was convenient for him because he wasn't too close, and yet he was close enough to get a good meal occasionally. Elsa had a tendency to scold him and mother him a little too much for his taste.

PATH TO THE GENERAL THEORY

Einstein missed his boys after they left. He tried to make arrangements to visit them in June, but he ran into problems with Hans. Hans announced that he didn't want to see his father and didn't want to go hiking with him. Einstein was heartbroken and blamed Mileva for turning the boys against him. He continued to write them letters, encouraging them in their homework and in music. Hans was now taking piano lessons. "Today, I'm sending off some toys to you and Tete. Don't neglect your piano, my Adu; you don't know how much pleasure you can give to others, as well as to yourself, when you play music nicely."[20] Einstein went on to tell Hans to brush his teeth every day. This is a little ironic in that when Elsa gave him a toothbrush he refused to use it. In another letter, "I will try to be together with you for a month every year so that you will have a father who is close to you and can love you."[21]

Einstein had hoped to go hiking with Hans in June, but with his refusal to go on the hike, Einstein decided to go to Göttingen where he met the mathematicians David Hilbert and Felix Klein. Hilbert was generally acknowledged as the greatest mathematician in the world, with the possible exception of Henri Poincaré. Einstein gave six, two-hour lectures on the latest developments in his extension of special relativity. He was particularly pleased that Hilbert understood everything in detail and was excited about the theory. "With Hilbert I was enraptured. He is convinced of the general relativity," he wrote.[22]

Arrangements were finally made in September for the hike with Hans. It was the last holiday Einstein would have before his intense burst of activity in the following months. They hiked in southern Germany, staying overnight at several inns, and they also took a boat trip on the Danube. Einstein was pleased with the success of the trip but embarrassed by Hans's questions. He kept asking if the family was going to get back together and when they would be coming to Berlin. Einstein was evasive and told him he preferred to have him educated in Switzerland.

When Einstein got back, he attacked the relativity problem with renewed vigor. The next couple of months were a critical time. His fervor for his new theory was intense, and he was determined to make it succeed. Skipping meals, he frequently worked far into the night. Day after day he sat glued to his desk, working as he had never worked before in his life. He began by taking a serious look at the noncovariant theory that he had developed with Grossmann. He had always been hesitant about it. Checking, he saw that it didn't predict the proper correction to Mercury's orbit, and in the case of the rotating disk, it didn't give the proper gravitational field. He finally decided to abandon it, and soon was looking again at the covariant theory he and Grossmann had devised two years earlier. Earlier he had found that it didn't go over to Newton's theory in the limit of weak fields. Rechecking, he found he had made a mistake. It did go over to Newton's theory. He could hardly believe his eyes and was ecstatic.

Beginning on November 4, 1915, he gave a series of four lectures to the Prussian Academy. On November 18 he showed that his new theory explained the anomaly in Mercury's orbit almost exactly. Furthermore, no factors had to be adjusted. And by November 25 the entire theory was complete. Einstein referred to it as "the greatest satisfaction of his life."[23] Elated and overwhelmed, he immediately wrote letters to his colleagues about his new theory.

"Imagine my joy over the practicability of general covariance and over the result that the equations correctly yielded the perihelion movement of Mercury," he wrote to Ehrenfest.[24] He finished the letter with, "For some days I was beyond myself with excitement." To some Dutch friends he said that the discovery had set off "palpitations," and he was sure something inside of him had burst. To Zangger, he wrote, "The theory is of incomparable beauty,"[25] and to Sommerfeld, "Make sure you have a good look at [the equations]; they are the most valuable discovery of my life."[26] To Besso, he wrote, "*My boldest dreams have come true*."[27]

The new theory of gravity was a complete departure from Newton's theory. Gravity was no longer a strange action-at-a-distance force; it was a curvature of space-time. Matter curved the space around it. The matter of our Sun, for example, curved the space around it, and when other matter passed through this curved space, it followed a "geodesic," in other words, the shortest path. Earth, in going around the Sun, was following the curvature of the space around the Sun. Gravity was no longer a force holding Earth to the Sun; it was curved space-time around the Sun. The deflection of light beams passing near the disk of the Sun had still not been verified, but there was an important change in relation to it. According to the new equations, it was double what Einstein had previously predicted. Another exciting prediction of the new theory was a "dilation," or slowing of time in an increased gravitational field. As a gravitational field increased in strength, a clock within it would run slower and slower. This meant that a clock at the top of a skyscraper would run slightly faster than one on the ground floor; the difference, as it turned out, was insignificant, however—only billionths of a second in a month.

For Einstein, however, a dark cloud appeared soon after he arrived at his theory. After he lectured at Göttingen, Hilbert became very excited about the theory. He knew little physics but was a first-rate mathematician and started looking for covariant equations as Einstein did. And indeed he found the same equations as Einstein. There were problems, however, with Hilbert's discovery. Einstein had gone back to the Ricci tensor after he discovered that it gave Newton's equations in the limit of weak gravitational fields, but he found that the equation that he and Grossmann had discovered earlier was satisfactory only to outside matter. Furthermore, it did not satisfy conservation laws. Within a short time, however, Einstein discovered that he could add a small term to the Ricci tensor, and everything came out all right. Einstein presented his equations to the Prussian Academy on November 25, 1915. A few days earlier, however, Hilbert had apparently presented the same equation, with the correction term, to the Göttingen Academy. Hilbert's paper was published in March 1916. John Stachel, of the Center for Einstein Studies at Boston University, discovered in checking Hilbert's manuscript, however, that he had made a change to his equation in late December. Apparently after seeing Einstein's correction term, he added it to his equation, and it was published in March with the submission date of November 20, 1915.[28]

Einstein was outraged; he was sure that Hilbert had plagiarized him, and he made no bones about it. Hilbert later wrote a letter to Einstein apologizing for not giving him credit and acknowledging that Einstein had made the discovery first. With this admission, the friction between the two men eventually disappeared.

The new theory was soon heralded as one of the greatest triumphs of mankind. But there was still one check that was needed: the bending of light around the Sun, which could be seen only during an eclipse.

A further joy for Einstein was soon to come. An astronomer, Karl Schwarzschild, received a copy of Einstein's new theory while he was at the front lines. Although he was director of the Astrophysical Observatory in Potsdam, he had volunteered for the war and was now in Siberia. With shells bursting around him, he read Einstein's paper and began working on a solution to his equations. Einstein had not obtained the most general solution, only a special case, namely the weak field solution. Despite the extreme conditions around him, Schwarzschild found a solution for the curvature of space-time around a massive spherical object such as a star, and he sent it to Einstein.

Einstein was surprised that someone had solved his equation so fast. He wrote back, saying, "I have read your letter with utmost interest. I had not expected that one could formulate the exact solution of the problem in such a simple way."[29] A few days later Einstein communicated the solution to the Prussian Academy. Schwarzschild had obtained the "exterior" solution of the equation; in other words, the solution for the region outside the mass (the star). But there was still the problem of the region inside the star. Schwarzschild immediately went to work on this, and within weeks he had a solution. Again, he sent it to Einstein, who presented it to the Prussian Academy. Schwarzschild's health had been declining for months, and within a short time he became very weak (he had contracted a rare metabolic disease). He was brought back to Berlin, where he died in May 1916. He was forty-one.

MILEVA'S BREAKDOWN

Einstein was on cloud nine after his breakthrough to general relativity. He was feeling so good that he thought it was time to make a complete break from Mileva, and in February 1916 he asked her for a divorce. Mileva was devastated. Separation was bad enough, and she wasn't ready for divorce; strangely, she was still hoping for a reconciliation. Einstein didn't find out how she felt until he came to Zurich a few weeks later to see the boys. The visit went well at first; he took Hans for a brief hike, but when he got back, he mentioned the divorce to Mileva, and a violent fight ensued with her refusing to let him see the boys anymore. Einstein left Zurich in a huff, thoroughly annoyed.

It was too much for Mileva. Shortly after Einstein left, she had a nervous breakdown. Doctors were not sure what was wrong with her, but it appeared that in addition to severe depression she had several small heart attacks. She was only forty-one at the time, but in many ways she was old for her age. She was too sick to look after the boys, and Besso took them. Later they went to Helene's, who was now living at Lausanne.

Einstein was sure at first that she was bluffing in a effort to avoid a divorce. Furthermore, within a short time, Pauline had her say about the matter. Writing to Elsa, she said that Mileva "only gets sick when it is convenient for her."[30] But the illness lingered, and Besso eventually convinced Einstein that it was genuine. Heinrich Zangger, who was a physician, examined her and thought that she had tuberculosis of the brain. At first, Einstein thought he should come to Zurich; he worried about the boys. But when he thought it over, he decided against it. He knew that Mileva would demand to see him and another fight would ensue, which would not be good for her.

Both Besso and Zangger were very sympathetic to Mileva, and both visited her frequently in the hospital. Einstein worried that his friends would have a poor view of him, and he wrote letters to them trying to justify his actions. He explained that he could no longer live with Mileva and that separation was inevitable, and he hoped that they would understand. Besso thought that Einstein mistreated her, but for the most part, he kept his views to himself. His wife, Anna, however, was outraged by Einstein's actions. As Marie Winteler's sister, she knew how Einstein had treated Marie, and she didn't mince any words. She added a postscript to one of Besso's letters telling Einstein exactly what she thought of him, and she didn't hold anything back. Einstein didn't notice that the writing was different and thought that Besso had written it. Shocked and worried, he immediately wrote back to Besso.

"For twenty years we have understood each other well," he wrote.[31] "And now I see an anger against me in you, on account of a female who is none of your concern. Fight it! She would not be worth it if she were a hundred times in the right!"

Einstein had barely put the letter in the mail when he realized it was Anna who had written the last few sentences in the letter. Embarrassed, he immediately wrote to Besso and apologized.

Not only was Einstein worried about what his friends thought about him, but he was equally worried about his boys. He had a strong love for both of them, but in many ways was closer to Hans. Hans was so much like him. He was good in school, but not outstanding, and he was a little rebellious; furthermore, he enjoyed music. All in all he was generally well rounded. But as Eduard grew older, it became evident that he was different. He was reading newspapers well before he went to school, and it soon became obvious that he had a photographic memory and could memorize long passages with ease. In first grade he was reading Goethe and Schiller, and when he took up the piano, he made incredible progress. His apparent genius scared both Einstein and Mileva; both felt that he should be held back and that something was wrong. One of the things that scared them the most was his intense mood swings. Einstein blamed part of it on the intense mothering he had received from Mileva because he was sick so often.

Fig. 18: Einstein at about forty years old

Mileva's recovery was slow, but she gradually began to get better. She had a serious relapse, however, when Hans got a letter from his father. She asked to see it, and when he refused, she went into a rage. No sooner had she begun to recover, however, than Einstein began to get sick. He had not eaten properly during his intense research and breakthrough to general relativity. Coupled with his family problems and the pressure from the war, everything began to overwhelm him. In a period of two months, he lost fifty-six pounds. He developed tremendous pains in his stomach and was sure he was going to die.

Chapter 13

Confirmation and a Passion for Determinacy

Einstein suffered in silence as his stomach problems increased, but finally the pain became so great, he had to do something. He told Erwin Freundlich, who urged him to see a doctor. Einstein was wary of doctors, but reluctantly took his advice. The diagnosis was a liver ailment. It may have been a relief to Einstein that he didn't have cancer, and would live, but the pain didn't go away for some time. Indeed, over the next few months, he would be diagnosed with several other problems, and his condition would not improve significantly.

Einstein was only thirty-eight, but he had neglected his health, particularly during the intense research that led to his breakthrough to general relativity. He frequently forgot to eat, and when he did, he ate what was at hand, cooking everything together in the same pot. And his sleeping habits weren't much better; he slept only when he was exhausted. In the end, it took a toll on him, and he would battle health problems for the next four years. His doctors urged him to go to a spa to rest, but Einstein had little faith in spas; he was sure it would do him little good.

Part of the problem was his lack of concern for his health. Asked if he worried about dying, Einstein replied that he didn't care. His theory of relativity was now complete, and that's all that really mattered, as far as he was concerned. He had pushed himself relentlessly over the previous few years. During 1916, for example, he had published ten scientific papers and written a book on relativity for the layman that is still in print today.[1] Most scientists are happy if they manage to publish a couple of papers in a year. Fueled by adrenaline, Einstein had obviously become a workaholic. Along with that he had marital problems, worry about his boys, and stress from the war. It's no wonder that he had a physical breakdown.

As his health deteriorated, he came increasingly under the care of Elsa. In the summer of 1917 he moved to an apartment next door to hers, so she could look after him better. He was put on a strict diet by his doctors, and Elsa prepared his meals. With the war on, it was difficult to get many of the things that were prescribed for him, particularly vegetables, but Elsa was resourceful and managed to keep him well supplied with the proper food over the next few years. Einstein praised her cooking and care to many of his friends. Indeed, his main passion as far as Elsa was concerned was her cooking. He was sick for several years during their early relationship, and this may account for the apparent lack of physical passion between them. The letters that he wrote to Elsa, for the most part, lacked the passion of those he had written earlier to Marie and Mileva, which were full of pronouncements of his devotion, love, and longings. More than anything, Einstein felt that he owed something to Elsa. As in the case of Mileva, when it came down to it, he had reservations about getting married, but felt obligated.

THE NEW UNIVERSE

About the time Einstein's health started to deteriorate, he was starting to think about how his theory of general relativity could be applied to the universe. Cosmology—the study of the universe—was not a well-developed science at the time. The accepted theory, namely Newton's theory, was over two hundred years old, and it had some serious flaws. It did not address such problems as: Does the universe have a boundary? And if so, where does it end? How is the matter in it distributed? Indeed, what does matter on a very large scale look like? Today we know that one of the most basic structures of the universe is the galaxy, which is a large conglomeration of stars. At that time there had been the suggestion that galaxies might exist, but nothing had been proved, so it was difficult to address some of the more significant problems.

When Einstein applied his newly discovered theory to the universe, he got a surprise. He found that the universe was unstable: it either expanded or collapsed. And since there was no evidence at the time that this was the case, Einstein decided that he would have to make some adjustments to his equations. The only way around the problem seemed to be to add a term, which he called the "cosmological constant." It would have no effect on a small scale (of the order of the solar system) but would make a difference on the scale of the universe. With this term, Einstein was able to stabilize his model of the universe. Furthermore, since matter curved space, Einstein knew that if he had a value for the average density of matter in the universe, he would be able to calculate its curvature. He got in touch with Edwin Hubble of the Mt. Wilson Observatory in California, who, as it happened, was working on this number and was able to give him an approximate value. Einstein substituted it into his equations and showed that the universe had a radius of approximately 10^7 light years.

A Dutch astronomer, Willem de Sitter, took a particular interest in Einstein's cosmological model. Holland was one of the few countries outside Germany that had access to German science during the war. De Sitter received a copy of Einstein's general theory of relativity and passed it on to the secretary of the Royal Astronomical Society in England, Arthur Eddington. Eddington was soon enchanted with the theory and asked de Sitter to write a summary of it for the *Monthly Notices* of the Royal Astronomical Society. Over the next few months, de Sitter wrote three articles. In the third paper, he discussed Einstein's new model of the universe, but while he was examining the model, he noticed that Einstein had missed one of the solutions to his equations. It was a strange solution, but it was a valid one, and de Sitter developed it and mentioned it in his third paper. To a first approximation, it gave a universe with no matter in it—an empty universe. This didn't bother de Sitter, since he knew that our universe was very close to being empty (the average density of matter in it is very low—about six hydrogen atoms in each cubic meter).

De Sitter's model may have been a bit odd, but it made an important predication. According to it, any two objects in the universe—two galaxies, for example—would repel one another. This meant that the universe would expand. And within a few years, Edwin Hubble would show that the universe was, indeed, expanding. Einstein had little faith in de Sitter's model. After all, without the cosmological constant his model also expanded, and he had added the constant to get rid of the expansion. But many people took de Sitter's model seriously, and for many years there were two models of the universe, and no one knew for sure which one was correct.

STOMACH PROBLEMS CONTINUE

Despite his strict diet, Einstein's stomach problems continued, and in March doctors discovered that he had gallstones. Adding to Einstein's misery was the uncertainty about his boys. Mileva was still sick and couldn't look after them. Eduard had been placed in a sanitarium, and Hans was at Zangger's. Einstein wanted a more permanent place for Hans. He knew he couldn't look after him, but Maja was now married to Paul Winteler, and they were living in Lucerne. Einstein was hesitant about asking her to take Hans, but he was sure it was the best place for him. He wrote to Besso to approach them. They agreed to take him, but for some unknown reason, nothing ever came of it.

In the late summer Einstein traveled to Zurich and stayed with Besso for a while, then traveled on to Maja's. He wrote to them both after he returned to Berlin, thanking them for their hospitality. To Besso and Anna he wrote, "You looked after me with such loving care I don't know how to recount it." In the fall he became director of the Kaiser Wilhelm Institute of Physics, which was part of his contract of 1914. At this stage it wasn't much of an institute, but it had a lofty goal, which was to encourage and foster research in physics. Haber offered him space at the university, but Einstein preferred to work at home. He needed a secretary to deal with the correspondence, so he hired Elsa's daughter Ilse, who was about twenty-two.

The war raged on, and although it didn't hit Berlin directly, it had a serious effect. Inflation caused the German mark to decrease in value against the Swiss franc, which caused problems in his payments to Mileva. Eduard was still in a sanitarium, and expenses were starting to mount. Einstein finally wrote to Zangger telling him that he could no longer afford to pay for him. But Mileva was still in and out of the hospital and could not look after him, so Einstein didn't have a choice.

Near the end of 1919, Mileva wrote to her mother asking if she would come to help, but she couldn't, so she sent Mileva's younger sister, Zorka. At first Zorka was of considerable help, but by February 1919 she had become depressed and had developed serious mental problems. Zangger had to put her in a sanitarium. He then hired a nurse to help Mileva, but she was expensive and Einstein worried that he couldn't pay for her. He wrote to Zangger telling him that he had sent 12,000 marks in 1917, which was his entire salary, and he could no longer afford to do that. (It makes one wonder what he had to live on.)

Late in the year, Einstein's health deteriorated again, and he was diagnosed with ulcers. Indeed, it's likely that he had had ulcers for several years. His stomach was acting up even while he was in college. Doctors told him that he would have to go to bed for several week's rest. Adding to his misery were problems with Hans. Earlier he had refused to write, and he was refusing again. Einstein was heartbroken, but finally in late January, he got a letter from him. Writing back, he said, " Your letter and postcard delighted me. Your concern about my illness was especially gratifying to me."[2] He had been in bed for a month by then, but the rest hadn't helped, and he was disappointed. He was thankful, however, that he could still work and correspond with his friends and colleagues from his bed.

ANOTHER DIVORCE ATTEMPT

It had been almost two years since Einstein had asked Mileva for a divorce, and it had caused so many problems that Einstein hadn't dared to raise the topic again. But soon Elsa was pressuring him again, and pressure was also coming from her parents. Einstein was unsure of how to handle it; he didn't want to cause more trouble, but Elsa had a reputation to worry about, since he was now spending considerable time in her apartment. She was also worried about her daughters. Ilse, the eldest, was twenty-two, and of marriageable age, and Elsa worried that rumors might be detrimental to her daughter's matrimonial prospects.

On January 31, Einstein wrote to Mileva asking for a divorce again.[3] He made her a generous offer: 9,000 marks per year instead of 6,000, and the proceedings from the Nobel Prize when he won it. It would be about 180,000 Swiss francs. He had, of course, not yet won the Nobel Prize, but he had been nominated every year (except two) since 1910, and it was almost a certainty that he would eventually get it.

Einstein braced himself as he waited for her reply, hoping that her reaction would not be the same as it was two years earlier. As it turned out, Mileva was in the midst of many difficulties and didn't have the strength to resist. Her sister Zorka had just been placed in a sanitarium, Eduard was still having physical problems, her brother had been captured by the Russians, and her health hadn't improved much. She conferred with Zangger and Besso, who finally convinced her that it was in her best interest to go through with the divorce. They told her that at his stage there was little chance for a reconciliation.

Mileva replied on February 6: "You will understand that with my current illness it is difficult for me to come to a decision. I do not have an overview of the situation—I must first accustom myself to the idea, also for the children's sake. I understand that you want an unhampered future; I don't know whether it is necessary for you and your work, but I don't want to stand in your way and obstruct your happiness."[4] She told him she had asked her lawyer, Dr. Zürcher, to look into what was needed for the divorce.

Einstein was relieved, but as it turned out, the proceedings would drag on for many months. There was, however, much less antagonism now between Einstein and Mileva, but the letter did lead to a problem. Shortly after Mileva received it, she talked to Anna Besso. Actually, she did more than talk; she poured out her heart to Anna, telling her all the wrongs that Einstein had precipitated on her and how he had been unfaithful as far back as 1912. Einstein found out through Zangger that Mileva was conferring with Anna, so he wrote to Anna, hoping she might be of some help in the divorce. He told her to think of things from his end: the problems arising from living near Elsa and her daughters. He ended the letter by asking her to put in a good word for him. "Make it clear to her how unkind it is to complicate the life of others."[5]

Anna, unfortunately, had no sympathy for him. Her response shocked and angered him. "I cannot sway Mileva anymore," she wrote.[6] "Debate it out with her." She then went on to tell him that she had no sympathy for Elsa. "If [she] had not intended to make herself so vulnerable, she ought not to have run after you so conspicuously. A mother with children ought to know what she is doing," she wrote. She then told him that his illness was no reason to marry her, and she reminded him that only a year or so earlier he had said, "Oh, as far as Elsa is concerned—you know, I really am not going to marry again."[7]

Einstein was so upset by the letter that he resolved never to have anything to do with Anna again. There was, of course, a problem, and Einstein was not sure how to deal with it. Besso was his best friend, and he didn't want to lose his friendship. He therefore wrote to Besso, mentioning that he bore no grudge against Anna. In an effort to smooth things over, he wrote, "No one is so close to me as you are and you know me so well and mean so well."[8] In the end the incident appeared to create no problems between the two men.

In May Einstein's liver acted up again, and he got jaundice. Again the doctors encouraged him to get some rest, and he finally agreed. He went with Elsa and her daughters to a resort on the Baltic Sea for two months. He invited Hans to come, but nothing came of it. Mileva certainly wouldn't have let Hans go if she knew Elsa was along.

Einstein and Mileva continued to write one another about the divorce. Despite the fact that there was an agreement between them, the proceedings dragged on. At one point Einstein said, "I wonder which will last longer—the war or our divorce proceedings."[9] As it turned out, the divorce proceedings did outlast the war.

In August Einstein was surprised by a letter from Zurich. The polytechnic and the University of Zurich had gotten together to make him an offer. As Einstein suspected, Zangger was behind it. Einstein was touched, but perplexed. The economy of Germany was terrible compared to that of Switzerland, and he had always had a soft spot for Zurich, but he now felt an obligation to the people of Berlin and was reluctant to leave. He was tempted and knew that if he accepted the offer, he would be closer to his children, but—unfortunately—also closer to Mileva.

Writing to Zangger, he said, "What do you want with an old wreck, an empty eggshell like me? What useful things I had thought up are alive in younger and fresher minds."[10] He suggested that Hermann Weyl would be a better choice. Still, he felt obligated and offered to spend two months lecturing at Zurich each year, and aside from expenses, he asked for no wages. This would allow him to see his boys each year, but not be too close to Mileva.

With the war still in progress, however, travel was difficult. Finally, there was a break on November 9, 1918, and it signaled the beginning of the end of the war. German sailors in Kiel refused to man their boats. The mutiny soon spread to the army, and soldiers refused to fight. The only fighting was in the streets of Germany with everyone demanding an end to the war. Germany was defeated; there was no hope. And indeed a surrender soon took place. The Kaiser abdicated and fled to the Netherlands. Einstein was delighted and quickly wrote to many of his friends with the news. It was the end of the war, but it wasn't the end of Germany's problems. Inflation had already been bad, but now it was rampant; the German mark continued to fall, and soon there was almost no way Einstein could pay Mileva the equivalent of 8,000 Swiss francs. Fortunately, Max Planck soon came to his rescue with a large raise.

THE ILSE AFFAIR

With the war over and his divorce now a sure thing, Einstein was relieved. It seemed that he would soon be marrying Elsa, but complications occurred. In many ways the situation at this point was similar to that after his father gave him permission to marry Mileva. Einstein had been courting Elsa for over six years, and she was now almost forty-three. He was approaching forty, and had been reluctant almost from the beginning to get married again. He enjoyed Elsa's culinary skills and felt grateful that she had looked after him so well for several years, but there was little passion between them. Indeed, it was almost as if Einstein tried to avoid passion. Furthermore, Elsa had a twenty-two-year-old daughter, Ilse, who in addition to being much younger, was also more attractive.

The problem began after it was clear that Einstein was getting a divorce. Ilse told a friend, Georg Nicholai, that Einstein would soon be marrying her mother, and to her surprise, Nicholai said that a marriage between her—Ilse—and Einstein would make a lot more sense. She could have children, and Elsa couldn't. This had to have shocked Ilse since she was infatuated with Nicolai at the time. But she didn't take it seriously. She was more surprised, however, when a few days later the topic came up at home. She wrote to Nicholai asking his advice, and despite the fact that she wrote "Please destroy this letter immediately after reading it" in capital letters across the top of the letter, it survived, and was found among Nicholai's possessions after his death.[11]

"Yesterday, the question was suddenly raised about whether Albert should marry Mama or me," she wrote. "This question, initially posed half in jest, became within a few minutes a serious matter which must be considered and discussed." She said that Einstein was prepared to marry either of them, and had told her that he loved her. She went on to say, "But he is far too decent and loves Mama too much. . . . You know how I stand with Albert. I love him very much; I have the greatest respect for him as a person. If ever there was true friendship and camaraderie between two beings, these are certainly my feelings for him." She emphasized, however, that she did not have "physical feelings" for him. Indeed, she said

Fig. 19: Ilse Einstein (later Kayser)

that she was "too used to regarding him as a father" to think of him as a husband. She stated that Einstein had not tried to persuade her in any way and said that he mentioned he would be happy if she merely lived in the same house as him.

One sentence is particularly interesting. It is: "I am absolutely not jealous of all the public glamour that would descend on Mama."[12] It is, of course, well known that Elsa loved the limelight, and Ilse must have realized it. The major thing that Ilse worried about if she did marry Einstein, was the reaction of her mother and the disappointment she would feel. "She would be disgraced," she wrote, "and she has suffered enough sadness and nastiness in her life."

We don't know how Nicholai replied to the letter, or anything about further discussions within the Einstein household, but in the end, Einstein did not marry Ilse. His divorce from Mileva finally came through on February 14, 1919, just a month before his fortieth birthday. The divorce forbade him from marrying again for two years, but it was filed in Switzerland, and as far as he was concerned, it didn't apply to Germany. Einstein and Elsa were married on June 2 in a quiet, simple civil ceremony. Indeed, Einstein's second marriage was no more elaborate than his first. Of course, he had already been living with Elsa for over a year.

It's interesting that only a year later, in 1920, when Einstein was invited to lecture in Scandinavia, he wrote, "I would take only one of the women with me, either Elsa or Ilse. The latter is more suitable because she is healthier and more practical."[13] And indeed he did take Ilse rather than Elsa. Also, in regard to the Ilse affair, it wouldn't be the last time

Einstein fell for his secretary (Ilse was his secretary for several years). When Ilse married a few years later, Einstein hired a new secretary named Bette Neumann. She was the niece of a good friend, and like Ilse was many years younger than Einstein (she was twenty-three). According to Albrecht Fölsing, "Einstein fell violently in love with her . . . and unlike his two marriages, this relationship aroused emotions that profoundly touched him."[14] Elsa eventually found out about the affair, and although she was annoyed, she did nothing to end it. Einstein broke off the relationship in 1924 writing to her that he "must seek in the stars that which was denied him on earth."

There's no doubt that Einstein's marriage to Elsa was very platonic. They had separate bedrooms at opposite ends of the apartment almost from the beginning. She said that she couldn't sleep with him because he snored too much, and there's no indication that Einstein wanted to sleep with her, either.

ECLIPSE AND FAME

If Elsa wanted to glow in the limelight from Einstein's fame, she certainly married him at the right time. Although he was fairly well known in Germany at the time she married him, he was not famous elsewhere. But all that would suddenly change in a few months.

In the last months of World War I, Sir Frank Dyson, the Astronomer Royal of England, pointed out that Einstein's theory could be tested during the upcoming eclipse of May 1919. Eddington was eager to help, even though he had become such an avid proponent of the theory that he didn't think it needed verification.

Preparations were made while the war was still in progress, and Eddington was put in charge of one of the two planned expeditions. His expedition would go to a small island in the Gulf of Guinea called Principe, and A. C. D. Crommelin would head a second expedition, which would go to Sobrel in Brazil. Both expeditions set out many months before the eclipse so that adequate preparations could be made. The weather at Principe had been ideal for weeks before the eclipse, but when Eddington woke on the day of the eclipse, rain was pouring on his tent. He was devastated, but went ahead with his plans anyway, hoping that the skies would clear in time.

As hoped, just before the eclipse, the rain abated and the clouds began to break up. The sight of blue sky spurred Eddington on, and when the eclipse began, he started taking photographs as quickly as possible. Clouds continued to cover the Sun, but Eddington was so busy he hardly noticed. Finally, just before the eclipse ended, the image of the eclipsed Sun broke through the clouds. In all, sixteen plates were taken, but only the last two showed stars around the eclipsed Sun. Comparison plates of the same region of the sky had been taken earlier back in England, and Eddington was soon making measurements. Conditions were far from ideal, but he got a first crude estimate which showed that there was a deflection, and it was close to Einstein's prediction. More accurate measurements would be needed, but he would have to wait until he got back to England for them. In addition, there were the plates from Sobrel.

Einstein's prediction for the amount of deflection was 1.74 seconds of arc. But as it turned out, Newton had also made a prediction that a light beam would be deflected by a gravitational field, in an appendix of his book *Opticks*. He had not made a numerical prediction, but Eddington showed that it would have been about .87 second of arc, which was half Einstein's prediction. Because of this, many people in England thought of the controversy as one between the Englishman, Newton, and the German, Einstein (Einstein was, of course, not a German citizen).

In England, after considerable work, Eddington arrived at a deflection of 1.61 seconds of arc, with a possible error of 0.30 second of arc, for the stars on his plates. He now had to wait for the Sobrel plates. He had heard that they had been successful, and he could hardly wait until they got back to England. But as it turned out, he had a relatively long wait. They had to get a comparison photo, and had to wait for the dark of the Moon to get it. Seven plates had been obtained with a four-inch telescope, and the stars on them were clear and sharp; several plates had also been obtained using an astrograph (a solar telescope), but the stars in them were slightly fuzzy.

When the Sobrel group finally got back to England, Eddington assisted them in making the measurements. They measured the astrograph plates first, and to Eddington's dismay, they gave a deflection of .86 second of arc—almost exactly the value predicted by Newton's theory. The group then went on to the plates from the four-inch telescope. They gave a deflection of 1.98 seconds of arc, with an possible error of 0.12 second of arc. Eddington was confused, but after a careful analysis of the astrograph plates, he concluded that a heating of the heliostat (system of mirrors that directs the sun's image to the photographic plate) had caused an error, and he decided to throw them out. That left his 1.61 seconds of arc and the 1.98 seconds of arc from the Sobrel plates. Averaging the two, he got 1.79, which was very close to Einstein's prediction of 1.74.

Meanwhile in Berlin, Einstein waited in anticipation. He expected some word by September, but heard nothing. Finally, he wrote to Ehrenfest in Holland, asking if he had heard anything. The answer was encouraging, but a little confusing. "Eddington found star dislocation at [the] Sun's rim provisionally . . . between .9 and 1.8 seconds of arc."[15] This covered both his and Newton's predictions. Einstein was still confident, however, that in the final analysis, his theory would be vindicated.

On November 6, 1919, a joint meeting of the Royal Society and the Royal Astronomical Society was held in London. There was only one topic on the agenda—the results of the two eclipse expeditions—and the meeting was buzzing with anticipation. According to Alfred Whitehead, "There was an atmosphere of tense interest that was exactly that of a Greek drama."[16] Sir Frank Dyson opened the proceedings with the words, "After careful study of the photos I am prepared to say that there can be no doubt that they confirm Einstein's predictions."[17] Further details of the expeditions were given by Eddington and Crommelin, and then the president of the Royal society, J. J. Thomson, said, "This is one of the most important results obtained in connection with the theory of gravity since Newton's day. It is one of the highest achievements of human thought."[18]

Einstein continued to wait for news, and it soon reached him. He was now at Leyden at the invitation of Paul Ehrenfest. While there, Ejnar Hertzprung showed him a letter from Eddington stating that the final result was 1.79 seconds of arc, confirming his theory. Two nights later the Dutch Royal Academy met, and with Einstein on the stage, the announcement was made. A roar of approval went up throughout the hall.

On November 7 the headlines in the *London Times* were "Revolution in Science—New Theory of the Universe—Newton's Ideas Overthrown." Within days, news reached New York, and a similar article appeared in the *New York Times*. Over the next few months, hundreds of articles appeared in papers across America and around the world. Einstein began receiving telegrams of congratulations, and letters began flowing in. Einstein was already well known in Germany, but now he was a worldwide celebrity. Within months he would become the most famous scientist in the world. And surprisingly, his fame would not fade; it would remain for the rest of his life.

ANOTHER DEATH IN THE FAMILY

Einstein was overwhelmed by the publicity, and also somewhat dismayed by it. People began to look upon him in awe, as if he were a superbeing. It was, at times, quite embarrassing for him, but with his usual sense of humor, he laughed about it and took it all in stride. Indeed, he posed for photographers so often that when a stranger, who didn't recognize him, asked him what he did for a living, he said, "I'm an artist's model."

One of the first persons Einstein informed when he heard his prediction had been verified was his mother. She now had stomach cancer and was staying with Maja in Lucerne. He knew that she was full of pride in his achievements, and he kept her informed on the latest developments. Pauline's cancer was inoperable, and it was soon obvious that she was going to die. She expressed the wish to spend her remaining days with her son, so Maja brought her to Einstein's apartment in Berlin. He had been married only for a few months at the time. She was in considerable pain, so the doctors gave her morphine, which caused her mind to wander and also to lose contact with reality at times. Einstein was distraught by her condition and tried to make her as comfortable as possible. Elsa also helped.

Pauline died near the end of February 1920. Einstein took her death hard. Earlier he had told Erwin Freundlich that no one's death would disturb him, and Freundlich's wife thought he was callous. But now, after seeing Einstein's tears, she knew he was softer than he pretended.

Einstein's feelings toward his mother had been ambivalent for some time. He had strong ties to her and would not go against her wishes. As we saw earlier, she was against his marriage to Mileva, and he delayed the marriage because of her. Many things about her annoyed him, but he knew that she was at least partially responsible for his success. She was determined to get him into the polytechnic, despite being underage, and she encouraged him. She also had tremendous pride in his musical ability. Her strong-mindedness and stubbornness bothered him at times, but he knew he could be just as stubborn as she was. Her death did not affect him as much as his father's death, but there's no doubt that he was distraught for some time and felt the loss deeply. In early March he wrote to Zangger, "My mother has died. . . . We are all completely exhausted. One feels in one's bones the significance of blood ties."[19]

BACKLASH

With the end of the war, Germany was overcome with inflation. The German mark dropped to an all-time low, and unemployment was rampant. Many Germans reached out to blame someone for their difficulties, and much of the blame was heaped on the Jews. Einstein, as a Jew, felt the wave of anti-Semitism; indeed, much of it was directed at him since he was now the best-known Jew in Germany.

Shortly after his return from Scandinavia in 1920, a public rally was held at the Berlin Philharmonic Hall. Paul Weyland, an engineer with political aspirations and a shady past, had organized a group he called "German Scientists for the Preservation of Pure Science." Einstein and others referred to it as the "anti-relativity group." Weyland managed to recruit physicists Ernst Gehrke and Philipp Lenard to his cause. Both men were openly anti-Semitic and did not hide their views.

Surprisingly, Einstein went along with Walther Nernst to the meeting, which was held primarily to blast his theories. Weyland began the meeting by denouncing relativity as a publicity stunt, and went on to call Einstein a plagiarist. Gehrke reported on a paper by Paul Gerber that had been published at the turn of the century. Gerber had tried to explain the anomaly in the perihelion of the planet Mercury, but his method had been unanimously rejected by astronomers. Gehrke, of course, did not mention this; he merely accused Einstein of stealing his ideas. Although Lenard did not speak, a pamphlet of his denouncing relativity was distributed in the foyer. Einstein did not appear to be perturbed by the proceedings. He even cheered and clapped along with the crowd mockingly when his theories were torn apart by speaker after speaker.

Many of Einstein's friends, however, were outraged by the allegations, and Laue, Nernst, and Rubens published a statement in the paper a few days later condemning the rejection of Einstein's theory and the malicious deformation of his character. On the outside Einstein appeared unruffled by the clamor, but inside he was seething, and he lost his temper. To the surprise of his friends, he sent a reply to his critics to the Berlin newspaper *Berliner Tageblatt*, and it was published on the front page the next day. He castigated both Gehrke and Lenard, saying that their statements were not worth a reply, nevertheless he was replying. He stated that Lenard had never achieved anything in theoretical physics, and his objections to relativity were superficial and made little sense. To many, he had gone too far; after all, Lenard was a Nobel laureate. Einstein claimed he had to defend himself, but later realized he had, indeed, gone too far and said a little sheepishly, "Everyone is entitled to one act of stupidity."[20] He resolved after that to never lose his temper again over such things.

At the end of September he was to encounter Lenard again, this time face-to-face. The Society of German Scientists and Physicians held their annual meeting at Bad Nauheim, and Einstein and Lenard were both in attendance. Max Planck chaired the physical science section. There was considerable tension before the meeting, and armed guards were hired to keep control and stop any violence.

The Einstein-Lenard debate started with Lenard stating that relativity "offends the common sense of a scientist."[21] He also objected to its elimination of the ether, which he felt was a very useful concept. Einstein countered by saying that what is viewed as common sense by one generation is looked upon as nonsense by another. He was jeered and shouted down by Lenard supporters, but he kept his temper. Planck was finally able to bring order to the assembly and ended the session.

Not long after Einstein sent his article to the Berlin newspaper, he shocked his friends again. He granted journalist Alexander Moszkowski an interview. Moszkowski asked him questions, not just about his theories, but on a variety of topics. Among them were his views on the possibility of life in space, hypnosis, and the role of women in science. It was the latter topic that got Einstein in hot water. He said he was strongly in favor of equal rights for women, and said they should have the same opportunities for education as men. But he didn't believe they were capable of becoming great scientists, and said that high achievement in science was generally beyond them. He emphasized, however, that Marie Curie was an exception.

When Max Born and his wife, Hedi, read a draft of the interview and heard that Moszkowski was going to publish it as a book, they became alarmed and urged Einstein to stop publication. He didn't need this kind of publicity, they said, and they worried about its effect. Einstein had already been called a "publicity monger," and this could be used to prove their case. Elsa, on the other hand, found the interview charming and urged Einstein to publish it. Einstein was caught in the middle; in the end he wrote to Moszkowski, telling him not to publish it, but it was published nevertheless. It came out in 1921 under the title *Einstein the Searcher: His Work Explained from Dialogues with Einstein.* Hedi Born was outraged, and when she heard that Elsa had urged publication, she wrote her an inflammatory letter. A feud between the wives ensued, and relations between Born and Einstein were strained for a while, but eventually everything was forgotten. The book, as it turned out, did little to tarnish Einstein's image.

THE NOBEL PRIZE

Einstein had been nominated for the Nobel Prize by Wilhelm Ostwald as early as 1910.[22] In fact, he was nominated every year after that with the exception of two years. Not only was he nominated frequently and by a large number of people, but he was nominated for several different achievements: special relativity, general relativity, Brownian movement, and the photoelectic effect. It's surprising that he could be nominated so frequently, yet go for so many years without receiving it. One of the major reasons, as it was later shown, was physicist Philipp Lenard. As a Nobel laureate, Lenard was on the nominating committee, and he had a surprising amount of influence on his fellow judges. Year after year Lenard protested that relativity had never been tested and was not worthy of the prize. He claimed it was of no value to science, and would eventually be proven to be incorrect, and therefore it would be an embarrassment to the Nobel committee if they awarded the prize for it.

By the early 1920s, however, it was obvious to the judges that they were going to have to award the prize to Einstein for something. There was still considerable controversy surrounding both the special and general theories of relativity, so one of the judges was assigned the duty of thoroughly studying the theories and giving a report. Unfortunately, the judge who was selected had little expertise in mathematics and physics and reported that he did not understand the theories. Einstein was therefore turned down again. In 1922, however, so many people nominated Einstein that the committee felt they had to give him the prize for something, so they selected his work on the photoelectric effect from 1905. Einstein was informed that he had won the prize while on a trip to Japan. No prize had been awarded in 1921, so although he received it in 1922, the prize was officially for 1921.

Einstein's feelings upon hearing that he received the prize are not known. It is known, however, that he kept a diary at this time, and surprisingly there is no mention of it in the diary. He had suspected for several years that he would eventually get it, so it was not a surprise. Furthermore, he never saw any of the money—it went directly to Mileva and the children.

QUANTUM THEORY

By the mid 1920s interest in relativity theory had begun to decline. The big attraction for physicists throughout Europe was a new discovery in quantum theory. Quantum mechanics, as the new theory was known, was a significant breakthrough—the first truly comprehensive theory of the atom—and it grabbed everybody's attention. It was a strange theory—different from previous theories in that it was based on probability and chance. Earlier classical theories had all been deterministic; in other words, everything could be calculated, or determined, exactly (at least in theory).

Einstein did not like the indeterminacy of the new theory. He refused to believe it was the final word. At best, he was sure, it was a temporary theory. His disdain for the new theory surprised many of his colleagues in that he was one of the few early supporters of quantum theory. Few paid any attention to Planck when he presented his idea of a "quantum" in 1900; indeed, even Planck himself thought that it was nothing more than a "temporary fix." But in 1905 Einstein used it to put forward the idea of light quanta, or photons, and with it he explained the photoelectric effect. Furthermore, in 1913 he was a strong supporter of Bohr's quantum theory of the atom, and in 1916 he used quantum theory to explain the emission and absorption of radiation. Over the years Einstein spent a lot of time thinking about quantum theory. At one point, in fact, he said, "I have thought a hundred times as much about quantum problems as I have about general relativity."[23]

In 1924 a young Indian physicist, S. N. Bose of the University of Dacca, sent Einstein a paper in which he had derived Planck's quantum formula by treating radiation as a gas made up of photons. Of particular importance, he counted the particles in a different way. Unlike Boltzmann, who had assumed that particles could be distinguished, Bose argued that they were indistinguishable. Einstein was enthralled with Bose's idea and extended it to atoms and molecules. Within a short time he had developed a new statistics for treating atoms and molecules. It is now called Bose-Einstein statistics, and many important problems have been solved using it.

Fig. 20: Einstein and his second wife, Elsa

The year 1924 also gave the scientific world another important breakthrough. Paul Ehrenfest of the University of Paris received a thesis from one of his students, Louis de Broglie, that confused him. De Broglie had extended Einstein's idea that light was both a wave and a particle to material particles. He suggested that all matter particles had a wave motion associated with them, and he determined the wavelength for a given particle of known mass. De Broglie was of noble lineage, and his brother was a well-known experimentalist, so his ideas could not be rejected outright. Ehrenfest therefore sent de Broglie's thesis to Einstein for his opinion. Einstein was enthusiastic and urged Ehrenfest to accept it. In fact, he was so enthusiastic about the suggestion that he urged experimentalists to look for evidence of the waves, and in 1927 they were found. De Broglie won the Nobel Prize for the idea a few years later.

Einstein was less receptive, however, to a paper he received from Max Born in 1925. One of Born's protégés, Werner Heisenberg, had developed a quantum theory based on strange "arrays" of numbers. What was particularly strange about these arrays was that if two of the numbers in them, say a and b, were multiplied in a certain order (e.g. $a \times b$), the result would not be the same as if they were multiplied in the opposite order ($b \times a$). Born eventually realized these arrays were used in mathematics, and were called "matrices." Einstein did not like the theory. "Heisenberg has laid a big quantum egg," he said.[24]

The following year another quantum theory was put forward, by Erwin Schrödinger of Zurich. His approach was based on differential equations, and equations of this type were used every day by physicists. They were much more familiar to physicists than matrices, and as a result, Schrödinger's theory became very popular within a short time. Einstein was enthusiastic about it—at least at first. There was, however, a problem: one of the central features of Schrödinger's theory was a wave function that he called ψ (psi). It somehow represented the wave properties of the problem, but Schrödinger was not sure exactly what it represented. Within a short time, Max Born of Göttingen showed that it (or more exactly, its square) was a measure of the probable position of a particle. In other words, the probability that a particle was at a particular position.

There were now two theories of atoms and molecules, and although they gave the same answer to all problems, they were based on completely different concepts. But Schrödinger soon showed that they were equivalent.

CLINGING TO DETERMINISM

In 1927 Heisenberg was able to explain some of the difficulties of the new theories with what he called the "uncertainty principle." According to it, you could not simultaneously determine the position and velocity (momentum) of a given particle. If you narrowed in on one and determined it accurately, the other would become "fuzzy" or uncertain. The same relationship applied to energy and time. About this same time, Bohr introduced his "principle of complementarity." He used it to explain the wave-particle duality of light. According to it, the particle aspect and wave aspects of light are not exclusive, but complementary. In short, one excluded the other, but both were needed to understand light.

As problem after problem fell to quantum mechanics, Einstein began to realize it was an important contribution to physics, and he nominated both Heisenberg and Schrödinger for the Nobel Prize. But it wasn't the theory's ability to solve problems that bothered him; it was the philosophical implications (which soon became known as the Copenhagen interpretation) that were repugnant to him.

Quantum mechanics was based on probability and chance, and, as such, it was indeterminate. Relativity and other classical theories, on the other hand, were determinant, in that you could calculate things exactly; probability was not involved. This was the aspect of the theory that Einstein disliked. He was sure it was only a temporary theory, and would eventually be shown to be an approximation to a more exact theory. For the rest of his life, he was convinced that a generalization of general relativity—a unified field theory—would eventually clear up the problems and give a deterministic view of the atomic world. Furthermore, he was sure that he would eventually find such a theory.

Although he had strong reservations against quantum mechanics, Einstein was full of admiration for the young "radicals" who developed the theory, namely Werner Heisenberg, Erwin Schrödinger, Wolfgang Pauli, and Paul Dirac. They reminded him of himself when he was young. "Quantum mechanics calls for a great deal of respect," he wrote.[25] "But some inner voice tells me it is not the true Jacob." One of the first chances that scientists had to hear Einstein's views came in 1927 at the Solvay conference in Brussels. He was asked to deliver a report on the status of quantum mechanics, but after initially accepting, he changed his mind, stating that he wasn't sure he was qualified to do it. Despite this, he had an overwhelming presence at the conference. Bohr was also there, and the two men had strongly different views. Bohr hoped to convince Einstein of the Copenhagen interpretation, but he had little success.

To many, it was the "battle of two Titans." Einstein would present a paradox, and Bohr, after considerable thought, would solve it and demonstrate that his interpretation was valid. This went on day after day, but there was no animosity between the two men. Indeed, according to all observers, the sessions were filled with humor and friendliness. Einstein had a lot of respect for Bohr. "He is truly a man of genius. . . . I have full confidence in his way of thinking," he had written to Ehrenfest only a few years earlier.[26] Bohr, in turn, had tremendous respect for Einstein.

Einstein went away from the 1927 Solvay conference slightly chagrined, but not defeated. He was still determined to show that determinism would eventually prevail and that there was a flaw in the basic tenets of quantum theory. And three years later he came well-prepared for the 1930 Solvay conference. He presented a paradox that threw Bohr completely off. He imagined a box filled with radiation that had a pinpoint shutter which would open and close by a clock. The box could be weighed before and after radiation emission, and the energy of the light could be determined exactly. Furthermore, the time could be recorded exactly, in conflict with the uncertainty principle.

Bohr spent a restless night thinking about it, but the next morning he had a solution. He pointed out that the weighing process gave rise to an uncertainty due to relativity theory, and as a result the uncertainty principle was valid. In short, he used Einstein's own theory to prove him wrong.

Einstein was embarrassed, but he still wasn't convinced. At one point Ehrenfest said to him, "Einstein . . . you are like your own critics of relativity."[27]

For several years Einstein and his theories continued to be attacked in Germany. And as the Nazi Party became increasingly powerful, Hitler became increasingly violent toward the Jews. Einstein, as one of the most prominent Jews in the country, began to worry. Soon it was obvious that his life was in danger.

Chapter 14

An Obsession for Unity

Einstein was worried, and he had reason to be. On June 24, 1922, his friend foreign minister Walther Rathenau was shot down in the street. Rathenau was a Jew, and Einstein had spent considerable time talking to him about politics, anti-Semitism, and Germany's future. Indeed, he had even warned him against taking the position of foreign minister. Einstein was sure that a Jew in such a powerful and conspicuous position would receive considerable antagonism in many circles. But Rathenau didn't listen to him.

Einstein canceled all his public speeches after Rathenau was killed, and he stayed out of sight. When an invitation to lecture in Japan came, he eagerly accepted, knowing that it would take him out of danger for a while. He enjoyed the trip and felt that it was a tremendous success; particularly pleasurable was his visit to Palestine in 1923 on the way home.

The following year Ilse married journalist Rudolf Kayser; she was twenty-seven. Einstein and Elsa both hated to see her move out of the apartment, but as it turned out, the newlyweds moved into an apartment only a short distance away. Kayser would write one of the first biographies of Einstein. Because Einstein had an aversion to anything personal being written about him, he carefully edited Kayser's version of his life, and as a result, much was left out. Einstein even forbid it from being published in Germany, so it was published in the United States.

In June 1924 Einstein also found out he was now a German citizen, which was a surprise to him. When he accepted the job in Berlin, he had told Planck and Nernst that he didn't want to become a German citizen, and they agreed that it was not a condition. But when Einstein won the Nobel Prize, the German committee at the Swedish ceremony thought he was German and was surprised to learn that he wasn't. They looked into it, and officials informed him that his election to the Prussian Academy of Science automatically gave him German citizenship. Einstein didn't object as long as they didn't strip him of his Swiss citizenship, which wasn't a requirement. He therefore now had both German and Swiss citizenships.

HANS GETS MARRIED

For years the relationship between Hans and his father fluctuated between warm and cold. Hans resented him for leaving the family, and it cast a shadow over his early life. At times, Hans was quite hostile toward his father. Einstein tried to make it up to him, but generally to no avail. Again and again Hans refused to write, but they did manage to have many enjoyable hours together, hiking in the mountains. Einstein was also disappointed when Hans went into engineering, but he soon accepted it, and when he heard that Hans was at the top of his class, he was pleased. Like his parents, Hans went to the polytechnic in Zurich, graduating in 1927 with a degree in civil engineering. He worked for a while for a steel company, but his real love was the mechanics of water and hydraulics.

Although they had had differences off and on for years, serious problems developed when Hans announced just before he was to graduate that he was going to get married.[1] He had met an older woman, Frieda Knecht, who lived in a nearby apartment. She was nine years older than Hans, and it was a shock even to Mileva when Hans announced he was going to marry her. Einstein nearly exploded when he heard the news. Frieda was not only much older than Hans, she was extremely short—almost dwarfish. Einstein did everything he could to stop the marriage. The irony of the situation was its similarity to his marriage to Mileva and his mother's reaction. Frieda was, in fact, very much like Mileva in appearance and personality, although she was Swiss, and Einstein's reaction to the marriage was every bit as vicious as his mother's had been years earlier. He called Frieda "a scheming older woman," and was sure that her "dwarfism" would be handed down to their children. Furthermore, he heard that Frieda's mother had undergone psychriatic treatment, so he had her investigated. As it turned out, she had had an extremely difficult life and had been depressed, but there were no serious mental problems. The possibility of mental problems worried Einstein, as Zorka (Mileva's sister) was now institutionalized, and Einstein feared that with mental problems on both sides of the family, their children would run a high risk of being mentally retarded.

Fig. 21: Einstein's oldest son, Hans, at approximately age twenty

Einstein even went as far as writing to Zangger and others in Zurich, asking them to try to reason with Hans. Strangely, even Mileva did not see the irony and similarity to her situation, and worried that it would not be a good marriage. But Hans was just as stubborn as his father, and nothing would change his mind. He got married on May 7, 1927. Einstein was annoyed and he still didn't give up. He implored Hans not to have any children. But again Hans did as he wanted, and two years later a son, Bernard, was born. And as Bernard grew up, it was soon evident that there were no problems. Furthermore, Einstein eventually had to admit, although reluctantly, that Hans's marriage was much happier than either of his.

HEALTH PROBLEMS

Einstein had already had his share of health problems, and early in 1928 he was struck down again. In February he collapsed after carrying a heavy suitcase for some distance uphill through the snow. The problem was soon diagnosed as a weak heart. He was treated by Berlin physician Janos Plesch, whom he had met earlier. Many people considered Plesch flashy and looked down on his association with celebrities. A number of doctors worried about the care he would give Einstein. But Plesch was up to the task; he put Einstein on a salt-free diet, forbade smoking, and told him he would have to rest. For the next four months, Einstein was therefore again in bed. The convalescence took longer than he hoped, but his heart was not in good shape, and if he had ignored the problem, he could have died. He was weak for almost a year.

Plesch reported that Einstein was an excellent patient, and didn't argue about the treatment. Elsa, however, soon found that his demands were driving her a little crazy. She was now acting not only as a nurse, but also as his secretary, and she didn't like it. With his heavy load of correspondence, it soon became obvious that he needed a full-time secretary. The sister of a friend of Elsa's, Helen Dukas, had just lost her job, and Elsa thought she should apply. Dukas, who was thirty-two at the time, was frightened at the thought of working for Einstein. She was well aware of his fame, and she was completely ignorant of science; nevertheless, with some trepidation she finally agreed to apply, and she was taken to Einstein's bedside to talk to him. To her relief, Einstein put her at ease immediately with his humor and easygoing style. She was hired and remained with him for twenty-seven years; indeed, she was soon treated as one of the family.[2]

By July Einstein felt well enough to go to the Baltic to continue his convalescence, but even as late as September, he wrote to a friend that he was still very weak. Being bedridden, however, didn't stop him from his work. In addition to his heavy load of correspondence, he continued his research from his bed. He now had an assistant, Dr. Walther Mayer. Einstein referred to him as his "calculator," since Mayer did most of the long, detailed calculations that were required in Einstein's research.

UNIFIED FIELD THEORY

Einstein's main interest at this time was an extension of his general theory of relativity. The success of general relativity had encouraged and pleased him, but there was another well-known field in nature that was not covered by the theory: the electromagnetic field. And the more Einstein thought about it, the more he became convinced that an extension of general relativity would include the electromagnetic field. It seemed natural; after all, there were too many similarities between the fields to ignore. Both depended on sources: matter was the source of the gravitational field, and charge was the source of the electric field (the overall electromagnetic field consisted of both an electric and a magnetic field). Furthermore, both fell off, or decreased, in the same way with distance, and both had an infinite range. Also, of particular significance, both (at least in theory) became infinite at the source.

There were, however, also significant differences in the two fields. There was a tremendous difference in their strength, with the electromagnetic field being 10^{37} times as strong as the gravitational field. Also, the gravitational field was always "attractive," and it was generated from only one type of source, namely matter. The electric field, on the other hand, was both repulsive and attractive, and it was generated from two types of sources—positive and negative charges. Another major difference was that electric fields could be shielded. Gravitational fields could not; nothing could stop the pull of gravity.

Einstein was encouraged by the similarities between the two fields, and he was sure he would be able to extend his general theory to cover the electromagnetic field. Actually, he hoped for more than this. The elementary particles—electrons and protons—had not yet been explained, and he hoped his new theory would explain them. What he really wanted, in more modern nomenclature, was a "theory of everything." There had been an attempt to explain the properties of particles using special relativity by Gustav Mie of Greifswald, Germany. But it had been shown to be flawed.

In 1919 Hermann Weyl, who was then at the University of Zurich, put forward a theory that was based on the idea of "parallel displacement," within curved space; this would cause a modification of the geometry of space-time, and Weyl made use of it. He sent his theory to Einstein, but Einstein soon found a flaw in it. The theory inspired him, but before he was able to publish anything, another unified field theory came into his hands. A mathematician, Theodor Kaluza of Königsberg, Germany, had written down Einstein's equations in five dimensions instead of four, and the extra dimension produced Maxwell's equations of electromagnetism. Einstein was amazed, more sure now than ever that the two fields could be unified. But Kaluza's theory gave no insight into the nature of elementary particles. Indeed, it seemed to give nothing more than Maxwell's equations, and as a result Einstein's enthusiasm for it soon faded.[3]

Einstein finally got into the act in 1925 when he extended general relativity by assuming it was nonsymmetric (a good example of a symmetric object is your body; it is symmetric on either side of a vertical center line). According to his idea, it had a symmetric and an antisymmetric part. General relativity was well known to be symmetric, so gravity was explained by the symmetric part. And to Einstein's delight, when he examined the antisymmetric equations, he found that they gave Maxwell's electromagnetic equations. He was sure he was on the right track, but as he examined the theory further, very little else seemed to come out of it, and he eventually had to abandon it.

In 1929 he went back to a variation of Weyl's parallel displacement theory and again was sure he was on the right track. As he got ready to publish, rumors leaked out that he had made a truly significant breakthrough. He had "tapped into the mind of God."[4] Hundreds of reporters assembled outside his house, hoping to interview him about his new theory. Knowing that the theory had not been tested or proven, he was embarrassed and tried to ignore them. His paper, however, had been presented to the Prussian Academy and was ready to be published, and to Einstein's dismay, it was released to the press. It was a relatively short paper, only six pages long, and it contained thirty-six equations, but the *New York Times* printed it, equations and all. No one understood it, or even knew if it was correct. Einstein therefore wrote a popular version of the paper and had it published in the *New York Times* and *London Times*.

All in all it was an embarrassment to Einstein, since a few months later he had to abandon the theory. Over the next few years, Einstein continued constructing and abandoning unified field theories. For a while he would be tremendously enthusiastic about a theory, then he would find a flaw, and the next day he would begin working on a new theory. Over the years he had many different collaborators, and he never gave up. Unification became an obsession with him, but it eventually became obvious that it was unlikely he would ever succeed. The major problem was the scope of the problem. Einstein wanted to unify the gravitational and electromagnetic fields, and hoped that the resulting theory would explain the electron and the proton. But over the years, two more fields of nature were discovered: the strong and weak nuclear fields. And the number of elementary particles skyrocketed into the hundreds. A simple extension of general relativity was not going to cover this. Furthermore, quantum mechanics was discovered in 1926, and it would somehow have to be incorporated into the new theory. Although Einstein struggled stubbornly for the rest of his life, he never achieved his goal, and indeed we still do not have a "theory of everything."

CAPUTH

The year 1929 was not only eventful as a result of the publication of a new unified theory, it was also the year of Einstein's fiftieth birthday. And as might be expected, there was considerable celebration. Several of Einstein's friends got together and chipped in to buy him a sailboat, and for several years it was his pride and joy. It was made of mahogany and had a small cabin and a toilet. With a sailboat, he needed a place to keep it—a summer home on a lake where he could use and enjoy it. Plesch therefore went to the Berlin city council and mayor and suggested that they give their "most famous citizen" a summer home as a birthday gift. They agreed, and after a brief search, they managed to locate a small house that was owned by the city. When Elsa went out to look at it, however, she found it was occupied. Although the city owned it, they had given a couple a long-term lease on it. Slightly embarrassed, the city council told Einstein to select some land and they would pay for it, but as Einstein and Elsa began their search, news of the gift leaked out, and several people were outraged and complained to the council.

When Einstein heard of the problem, he quickly declined the gift. He and Elsa had found a lot on a lake at Caputh, a small rural village near Berlin, and they decided to use their savings to build on it. It was an ideal setting with a lake in front, where Einstein could sail whenever he wanted. They built a house, and Einstein soon grew to love the place. Over the next few years, he spent as much time at it as possible. Hans was one of the first to come and go sailing with him. His sister, Maja, also sailed with him frequently. He loved the tranquillity, the quiet and peacefulness of the place, and was completely relaxed when he was at Caputh.

WOMAN TROUBLES

Einstein's assistant, Walther Mayer, also frequently came to Caputh, and he and Einstein continued working on a unified field theory. Although everything seemed to be peaceful on the surface, visitors soon noticed that there was considerable strain between Elsa and Einstein. They had many arguments, and the arguments were usually about one thing: women. Einstein always tried to keep his private life private, so no letters survive, but the live-in maid saw everything, and much of what we know of this period of Einstein's life comes from her. "Einstein loved beautiful women, and they loved him in return," she said.[5]

As a celebrity, Einstein was besieged by women everywhere he went, and there's no doubt that he loved the attention he got from them. Elsa put up with it dutifully, and tried to laugh it off, but in a few cases, things went a little too far for her. What particularly disturbed her was the brazenness of some of the women, who would call on Einstein and whisk him off in their car. One of the first was Toni Mendel, a rich Jewish widow with a chauffer-driven limousine. She would call on Einstein, bring Elsa chocolate creams (one of her favorites), then she and Einstein would disappear. They would go to concerts or the opera, and occasionally he would stay overnight at her plush villa at Wannsee. She paid for the tickets, but Einstein was expected to pay for some of the incidentals, so he would have to go to Elsa for pocket money. According to the maid, this is when the fireworks usually began. On several occasions Elsa refused to give him money to spend on his "hussy," and Einstein lost his temper. He even took Mendel sailing, and Elsa was particularly outraged when she found a low-cut bathing suit on his sailboat one day.

Estella Katzenellenbogen was another rich widow with a chauffer-driven limousine who also occasionally called on Einstein. She owned a chain of floral shops in Berlin. But the one that made Elsa see red was Margarete Lenbach. She was blond, Austrian, and much younger than Einstein, and what worried Elsa the most was that she was extremely attractive, almost beautiful. She came to Caputh every week with treats—usually pastries—for Elsa, and soon she and Einstein were off by themselves. Einstein's eyes really lit up when she visited, and there was a tremendous amount of laughing and joking. Elsa hated it and usually went to town when the "Austrian," as she called her, arrived.

Once after the "Austrian" had visited, the maid heard Elsa having an argument with her two daughters. As she cried on their shoulders, they told her she had only two options: put up with it, or divorce him. It was soon obvious what she had decided. The visits continued.

Ilse had now been married for several years, and it was no surprise that she got married. She always wore the latest fashions and easily mixed with society, but when her younger sister, Margot, announced that she was getting married, it had to have been a surprise. She was incredibly shy and would frequently hide when Einstein had visitors. On more than one occasion while Einstein was discussing something with a visitor, Margot was sitting under the table, hidden from view by a long tablecloth, waiting for the visitor to leave.

Margot married Dimitri Marianoff, a not-too-successful journalist. It later became known that he had ulterior motives in marrying her. He was particularly interested in doing a biography on Einstein but found that it was almost impossible to gain access to him, so he began dating Margot.

Fig. 22: Einstein's house at Caputh

Marianoff knew that Einstein disliked anything personal being written about him, and as a member of the family, he would be able to see exactly what was going on. And indeed when his biography came out, he did go into the personal life of Einstein in considerable detail, giving us his opinion of Einstein's view of women. Einstein, of course, was not happy with his biography. Marianoff eventually divorced Margot, telling her he could no longer support her. Margot never remarried, and stayed with Einstein for the rest of her life. Indeed, she became very close to him and was quite jealous when other people took up too much of his time.

TO AMERICA

In December 1930 Einstein set sail for America. He was accompanied by Elsa; his secretary, Helen Dukas; and his collaborator, Walther Mayer. Einstein wanted Mayer along because he hoped to get some time to work on his latest unified field theory during the voyage; for him it was to be a "working" vacation. He was headed to California, but the boat would

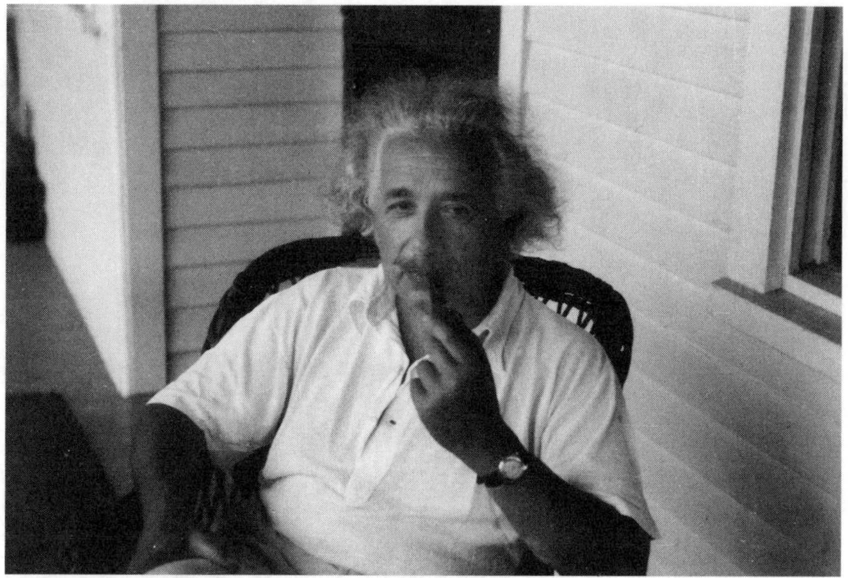

Fig. 23: Einstein at Caputh

dock in the New York harbor for five days. Einstein planned on staying aboard and working, but when the news got out that he was aboard, the boat was besieged by reporters. Everyone wanted to interview him, but he was embarrassed; he couldn't speak English and needed an interpreter. Furthermore, many of the questions that reporters asked him were not about science, and they bewildered him.

Einstein finally relented and went ashore, and for the next five days, he made numerous appearances, several of them for charitable causes. He was even made an honorary citizen of New York City. It overwhelmed him, and he was relieved when it was over. From New York the ship passed through the Panama Canal and arrived in San Diego on December 30, 1930, where again reporters and photographers besieged him.

During his stay in San Diego, he and Elsa were housed in what he called the "gingerbread cottage" on the California Institute of Technology (Caltech) campus. Robert Millikan was their host and guide. Einstein's main objective while at Caltech was to visit the large new telescopes on Mt. Wilson and to talk to Edwin Hubble and Richard Tolman. Hubble had

Fig. 24: Margot Einstein

just presented evidence for the expansion of the universe, and Tolman was the leading American authority on relativity and cosmology. Indeed, only a couple of years later, Tolman would write his classic book *Relativity, Thermodynamics, and Cosmology*. In it he covered both special and general relativity, and discussed Einstein and de Sitter's models of the universe in considerable detail.

Hubble invited Einstein to look through the hundred-inch Hooker telescope, and Tolman had lengthy discussions with him about cosmology. After talking with Tolman, Einstein announced that he was abandoning his cosmological constant. He now knew that if he had not introduced it, he could have predicted the expansion of the universe; he therefore called it the "biggest blunder of his life."[6]

At one point while visiting Hollywood, Millikan asked Einstein if there was anyone he would particularly like to meet. "Charlie Chaplin," replied Einstein. Chaplin was contacted, and he said he would be pleased to meet Einstein. They had lunch together, and Chaplin later invited Einstein to his estate. Einstein also attended the premier of his movie *City Lights* with him. He later said of him: "Just as in his films, Chaplin is an enchanting person."[7]

Einstein returned to Germany in March, but things had worsened while they were away, so he didn't stay long. He presented two papers to the Prussian Academy, one on cosmology and one on his latest unified field theory, then took off for England. He gave the Rhodes lecture at Oxford and received an honorary degree from the university. The month he spent in England was relaxing after his hectic schedule of earlier that year. His host was Fredrick Lindemann, and while he was there, Lindemann convinced him to come to Oxford for a month each year.

Einstein was back in Germany for the summer, spending most of his time at Caputh. But as the Nazi threat continued, Einstein became increasingly active in politics and began speaking out against the Nazis. He was outraged by the anti-Semitism and supported many pacifist and humanitarian causes, which later got him in trouble. Some of these pacifist and humanitarian causes were nothing more than fronts for the Communist Party, and as a result, Einstein was soon labeled a communist, which caused him some discomfort. He later made it clear that he was not a communist, saying, "[I] would like to state that I have never favored communism and do not favor it now."[8]

AN OFFER HE COULDN'T REFUSE

Einstein had a commitment back in the United States near the end of the year. The most important result of this trip was the initiation of an offer of a position at a newly forming academic institute in New Jersey. The new institute was the brainchild of Abraham Flexner, an educator who was determined to elevate the status of American education and assist scholars in their research. He had a five-million-dollar grant from some wealthy donors and was planing on building an institute where scholars and scientists would have no duties except research. It would be a place where they would be free to pursue their research, and there would be an emphasis on mathematical and theoretical research.

During their first meeting, Flexner did no more than ask Einstein's opinion of the project, and of course Einstein was strongly for it. Einstein met Flexner again at Oxford in the spring of 1932, and this time Flexner asked him if he would like to be part of the institute. In short, he made him an offer and said it would be "on his own terms."[9]

Einstein was interested, but despite the unrest in Germany, he was reluctant to leave. He had grown to love Caputh and felt an obligation to many of his German colleagues. He told Flexner, however, that he would consider a half-time position, spending half of the year in Princeton (where the institute was now going to be located) and half in Berlin. Einstein also told him that he wanted to take his collaborator, Mayer, with him, and that Mayer's position should be independent of his. Flexner was reluctant about Mayer, but said he would think about it. When the discussion got around to wages, Flexner asked Einstein what he thought a fair wage was. "Would $3,000 be okay?" said Einstein. Flexner laughed, thinking he was joking. "Could I get along on less?" said Einstein, when he heard Flexner's laugh.[10] Flexner realized he was serious and said he would talk to Elsa about how much they would need. A little later Flexner made an offer of $10,000 plus expenses for both him and Elsa.

Einstein was enthusiastic about the new job, but there were still problems, and Flexner visited him again at Caputh in the summer of 1932 to iron out the remaining difficulties. Anti-Semitism was now greater than ever in Germany, and Flexner was sure that Einstein would sign the final papers. The conversation between the two men lasted well into the night, and just when Flexner was sure he finally had him convinced, Einstein said he would think the offer over, but he wouldn't sign yet. He still couldn't bring himself to leave Caputh for so long, and there were still issues concerning Mayer.

Several of Einstein's friends were alarmed when they heard he was holding out. They encouraged him to sign, saying that Germany was no longer safe for him. Furthermore, Elsa was getting more frightened by the day, with the stories she was hearing. But Einstein was convinced that the Nazis were only a temporary problem and would eventually be voted out of power.

THE NAZI RISE TO POWER

The 1929 Wall Street crash and its reverberation around the world was what the Nazis were waiting for. Unemployment was rampant throughout Germany. Soup lines were lengthening, and despair among the people increased. This was exactly what Hitler wanted; he promised them the sky. He would get rid of unemployment, he would put food back on the table, and he would make Germany great again. The greatest problem, he emphasized, was the money-mongering Jews, and he would do something about them. People had to have something to believe in, and they soon turned to him.

Einstein still had a commitment to go to Caltech at the end of the year. Ironically, it wasn't only problems in Germany that plagued Einstein. When he went to get his visa, he found that he also had enemies in the United States. The Women's Patriot Corporation had originally been organized to help women get the vote, but it was now focusing on stopping "undesirables' from entering the United States, and at the top of their list of undesirables was Einstein. They claimed he was a communist, that his relativity theory was overblown, and that they were outraged by his pacifist views.

Einstein was shocked when he first heard about them, but he laughed it off. "Never have I experienced from the fair sex such energetic rejection of all advances; or if I have, never from so many at once," he said.[11] But the group had an effect. Einstein was interrogated for almost an hour as he tried to get his visa. This was a surprise to him since he had never been questioned before. Finally, with no visa in hand, Einstein left the consulate in anger, but he was quickly called back the following morning; an apology was given, and the visa was issued.

By now Einstein had become convinced that his time in Germany was limited, and he finally accepted Flexner's offer, but it would not begin until October 1933. On the surface this was only to be a "visit" to the United States, but as Einstein left Caputh he told Elsa to turn around. "Take a good look," he said. "It's the last time you'll ever see it."[12] And indeed it was.

Paul von Hindenburg had won the election in March, but barely, and he was now senile. Hitler saw his chance and took it. Under pressure from the Nazis and the communists, Hindenburg appointed Hitler chancellor of the Reich, and it wasn't long before Hitler took over and declared himself dictator. He seized power on January 30, 1933, while Einstein was in the United States. Einstein now knew for sure he would not be returning to Germany. Nevertheless, when his commitment was up at Caltech in March, he sailed back to Europe.

Once Hitler was in power, the Nazis didn't hold back, and Einstein was at the top of the list of hated figures. The press attacked him as a traitor, and as might be expected, Philipp Lenard and a number of other scientists joined in, again calling relativity a worthless contribution to science. Einstein fought back in the press, condemning Hitler whenever he had a chance, and calling for the civilized world to intervene against him.

Soon there was a price on Einstein's head in Germany. Hitler's brownshirt storm troopers raided his apartment in Berlin and his house at Caputh. Everything was confiscated, including his prized sailboat. His apartment was locked up. Then they went after his savings and confiscated 30,000 marks he had in Berlin banks. Ilse was still in the apartment in Berlin when they first raided it, and she was scared out of her wits as the storm troopers ransacked the apartment, taking everything they could. She had been trying to salvage Einstein's papers, and indeed she and her husband did manage to save most of them.

Interestingly, there was a report that the Nazis planned on taking Ilse and Margot as hostages in their effort to get at Einstein, but the plot failed when they managed to escape to Paris.

Einstein's ship docked in Belgium, and he and Elsa took up residence at Le Coq sur Mer, on the Belgian coast. The German consul in Belgium warned Einstein about attempting to go to Germany. "They will pull you through the streets by your hair," he said.[13] But Einstein had no wish to return. Nevertheless, he had important business with German officials to attend to. The first thing he did was reject his German citizenship. Then he wrote to Bernard Rust, the Nazi minister representing the Prussian Academy, and resigned from it. Strangely, there was some discussion of rejecting his request, then dismissing him from the academy. But it didn't happen. Planck and others were disturbed by the developments, but it was difficult for them to say anything.

When Mileva heard of Einstein's plight, and that there was a price on his head, she offered her apartment to him and Elsa. Einstein was touched by the offer, but he refused. Switzerland was a little too close to Germany for comfort. The threats on Einstein's life continued. A friend brought him a magazine from Germany that featured the "top enemies of the Nazi Regime." Einstein's picture was on page one, and below it were the words "Not yet hanged."[14]

Einstein was on good terms with the royal family of Belgium, and the queen sent several detectives to guard him. They weren't much of a consolation to Elsa, however; she worried so much she could hardly sleep at night. Einstein was merely annoyed, and a little bit embarrassed by all the fuss over him. Ilse, Margot, Dukas, and Mayer eventually joined them at Le Coq, but when an invitation came from Oxford, Elsa was anxious to get as far away from Germany as possible.

LAST VISIT WITH EDUARD

As they were preparing to leave for England, Einstein received word that Eduard had undergone a breakdown. He appeared to have schizophrenia, and Mileva had been unable to handle him. Despite the possible danger to him, Einstein rushed to Zurich.

The relationship between Einstein and Eduard had always been a bit strange. Einstein deeply loved both of his sons, but he worried about Eduard. There was no doubt that he was close to being a genius (although not in math and physics; he preferred literature, the arts, and music). His feelings toward his father were mixed. On the one hand, he revered and almost worshipped him, and he tried desperately to gain his acceptance. He wrote many strongly assertive letters to him trying to challenge him intellectually. Einstein didn't know what to make of them, and he complained to Hans that the letters were very impersonal. But above all, Eduard blamed Einstein for deserting his family, and he never forgave him for it.

Eduard breezed through high school, and by the time he was ready to go to university, he had decided on medicine. Rather ironically, his major interest was psychiatry. But he had barely started when he became involved with an older woman (also a medical student). When she rebuffed him, it sent him into despair, and many people think this is what triggered his schizophrenia. The first signs came when he began skipping classes and ignoring his friends. Mileva knew she had to take drastic measures, so she got in touch with Einstein.

Einstein was shocked by what he saw. By now Eduard's ambivalent feelings had become one-sided. He told his father that he hated him and that he had "cast a shadow over his life."[15] Einstein was heartbroken and wasn't sure what to do. Zangger was confident that Eduard could, with treatment, still resume a normal life. But Einstein was not so sure; to him, Eduard's problems were hereditary—from Mileva's side of the family—and as such, little could be done. He also blamed Mileva for mothering him too much. Nevertheless, he urged Mileva to get the best psychiatrist and doctors for him, saying he would pay for them. It was difficult, however, to treat Eduard. He had studied psychiatry extensively on his own, and when the doctors questioned him, he usually knew what they were looking for and how his answers would be interpreted.

Eduard was given insulin treatments to induce a short-lived coma. According to Hans, the thing that ruined him were the shock treatments. Electric probes were attached to his head, and large currents were discharged. (In some cases these treatments caused heart attacks, yet strangely psychiatrists continued to use them for years.)

Eduard was never the same again and spent most of the rest of his life in sanitariums. He was not violent, but he could not be left on his own. He would get lost if he ventured only a short distance from his home. A couple of years after his treatments, he, along with a male nurse, went to visit Maja. She was shocked by his appearance. He stayed for several days playing the piano and staring off into space.

When Einstein returned from Zurich, he was severely despondent over Eduard's condition. He already had more problems than he could handle, and this had to come on top of them all. He considered taking Eduard to the United States, but it didn't work out. When he left Zurich, he was seeing Eduard for the last time. He never saw him or Mileva again for the rest of his life.

LEAVING EUROPE FOREVER

Einstein went to Oxford and delivered several lectures. While there, he learned that his friend Paul Ehrenfest of Leyden had committed suicide. He had shot his mentally deranged son, then shot himself. Einstein had been very close to Ehrenfest; he had visited him frequently in Leyden, playing duets with him. Furthermore, he was with Ehrenfest when he heard of the results of the British eclipse expedition. He took his death hard.

In September Einstein, along with Elsa, Dukas, and Mayer, departed for the United States. This time they went directly to Princeton. The building for the Princeton Institute of Advance Study was not yet completed, so the institute was set up at Fine Hall on the Princeton University campus. Einstein and Elsa both grew to love Princeton and its peaceful surroundings. Einstein could now work on his unified field theory unhindered. But he had barely got down to work when Mayer announced that he was leaving. Einstein was dismayed; he had struggled so hard to get Mayer as part of his contract and had even threatened Flexner that he would not come without Mayer. And now Mayer was leaving. The reason, it is believed, is that he had lost faith in the unified field theory.

Einstein didn't have to worry, however. He soon had new collaborators; among them were Banesh Hoffmann, Leopold Infeld, Peter Bergmann, Valentine Bargmann, Boris Podolsky, and Nathan Rosen.

Fig. 25: Einstein's youngest son, Eduard, at approximately age fifteen

THE DEATH OF ILSE

The Einsteins had barely got settled in Princeton when Elsa got a letter from Margot. Ilse was deathly ill.[16] They had delayed telling her, not wanting to worry her, but it was now very serious, and it looked like she might die. Elsa was very close to her daughters and was anxious to get to Paris as soon as possible, but Einstein still had a price on his head in Europe, and as much as he loved Ilse, he was reluctant to go. Elsa therefore went alone. She arrived to find Ilse on her deathbed; she had lost so much weight and looked so pale, it scared Elsa. She had first been diagnosed with cancer, but they soon found out it was tuberculosis. But Ilse had rejected traditional treatment and put her faith in psychoanalysis. She died soon after Elsa arrived, and the shock was so great that Elsa never got over it. Ilse was thirty-seven.

With Ilse's ashes, and accompanied by Margot, Elsa headed back to the United States. One of her neighbors from Princeton recognized her on the boat and noticed how she had changed. When she asked what was wrong, Elsa broke down in tears, "It's my Ilse. She is dead. I cannot bear it," she sobbed.[17]

Elsa was heartbroken over her daughter's death, and she had lost considerable weight and looked terrible by the time she got back to Princeton. Ilse's death and Elsa's tremendous grief weighed heavily on Einstein. He sympathized with her, but continued to work on his research. Indeed, it was during this time that he managed to publish, along with two collaborators, Boris Podolsky and Nathan Rosen, a paper of fundamental importance. Referred to as EPR (the first letter of each of their last names), it pointed out a serious paradox in quantum mechanics, and for many years, was seriously considered by scientists.

MORE DEATHS

In September of the following year, 1935, Einstein heard of the death of his good friend Marcel Grossmann. Grossmann had had multiple sclerosis for several years, and had gradually wasted away. He had played an important role in Einstein's life. Grossmann had unselfishly loaned him his notes, which allowed him to pass the finals at the polytechnic. He had helped him get a job at the patent office when Einstein had been rejected by almost everyone, and he had helped him with the complicated mathematics when he was trying to formulate general relativity. Indeed, he and Grossmann had actually arrived at the correct equation without realizing it.

Einstein wrote to Grossmann's wife as soon as he heard the news. "I remember our student days. He, the irreproachable student, I myself, unorderly and a dreamer. He, on good terms with his teachers and understanding everything. I, a pariah, discontent and little loved. But we were good friends."[18]

As he grieved over Grossmann, another shock was on its way. Elsa developed a problem with one of her eyes. Einstein urged her to go to the doctor. The diagnosis was not good, and the doctor told only Einstein, who kept the news from her. They said she had serious heart and liver problems and wouldn't likely live long. She was quite surprised when Einstein became very attentive and loving. But her health continued to deteriorate. Hoffmann reported that he could frequently hear her cries as he and Einstein worked on their unified field theory. Elsa died on December 20, 1936.

Chapter 15

A Desire for World Peace

Einstein had been a strong pacifist for many years, even when he was quite young. There were no conditions on his feelings, and they came out in his writings and speeches. In 1921, for example, he wrote, "I would absolutely refuse direct or indirect war service and would try to persuade my friends to do the same, regardless of the cause of a war."[1] He continued in this vein through the early 1930s. In 1931 he wrote, "I believe serious progress [in the abolition of war] can be achieved only when men become organized on an international scale and refuse, as a body, to enter military or war service."[2]

Einstein gave many pacifist speeches in these years. One of his most famous and controversial was made at the Ritz-Carleton Hotel on December 14, 1930.[3] It was carried in the *New York Times* the following day, and many people were disturbed by it. Referred to as the "2 percent" speech, its thrust came in the statement, "If only 2 percent of those called up declared that they would not serve, and simultaneously demanded that all international conflicts be settled in a peaceful manner, governments would be helpless." According to Einstein, this 2 percent would amount to so many people there wouldn't be enough jails for them all, so they obviously couldn't jail them.

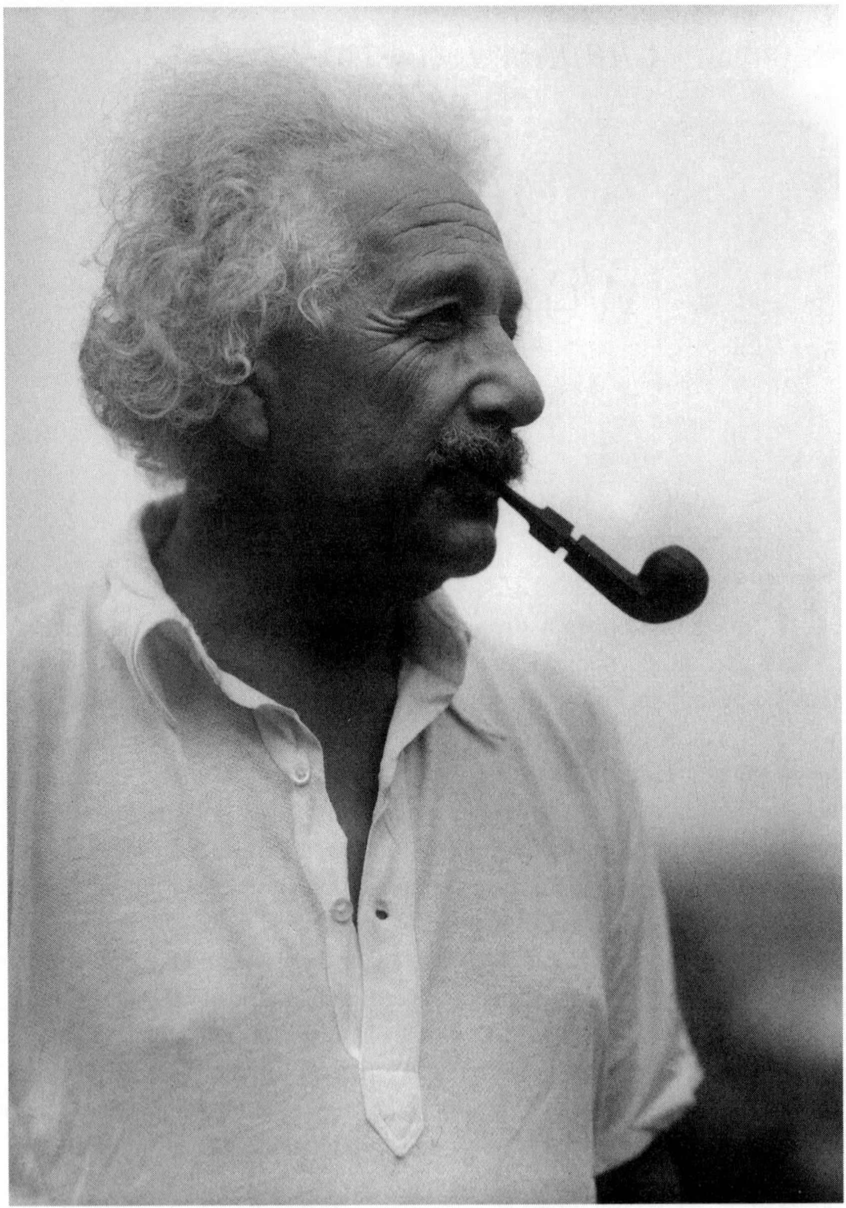

Fig. 26: Einstein enjoying his pipe

Einstein was on his way to California when he made the speech, and news of it reached his California host, Robert Millikan, before he got there. Millikan was not pleased, and worried about what Einstein might say while he was in California. But Einstein knew enough not to overstep his bounds, and there were no embarrassments for Millikan.

The rise of Hitler in 1933 changed Einstein. Although his pacifist views were still strong, he knew that something had to be done about Hitler, and that wouldn't be possible without military action. He began referring to himself as a "militant pacifist." In short, he was willing to fight for peace, even if it meant taking up arms. He startled many of his pacifist followers in 1933, when, shortly after he returned to Belgium, he announced that "under today's conditions, if I were a Belgian, I would not refuse military service, but gladly take it upon me in the knowledge of serving European civilization."[4]

In the years after 1933, Einstein's hatred and disgust for the Nazis and Hitler began to extend to the German people. He blamed them for the atrocities against the Jews, as much as he did Hitler. They had put Hitler in power, and continued to support him even after it was obvious how aggressive and cruel he was.

A CALL FOR HELP

For the first few years that Einstein was in the United States, he was busy assisting Jewish emigrants coming to his new country. Many were fleeing Europe because of Hitler's new policies, and Einstein was never one to turn down anyone's plea for help. His friends frequently said (jokingly): There are two things Einstein would never turn down. The first is an opportunity to make music, and the second is an urgent request from a poor or oppressed person. And indeed he helped many of his family members and their children, but he also helped many strangers.

In 1936 he encouraged his son Hans to come to the United States, and he arrived in 1937. Hans eventually ended up as a professor of civil engineering at the University of California at Berkeley, where he had a brilliant career in hydraulics and water control. Over the next few years, Einstein signed affidavits for many people, and he put $2,000 in a special account for each of them. His resources, however, were soon pressed to the limit, and he asked his good friend Leon Watters to help, and indeed Watters gave him considerable assistance. Einstein realized he could only do so much, so he tried to encourage organizations such as church groups, the Red Cross, and others to help, and he was also successful in this.

His sister, Maja, was one of those who had to flee Europe. She was living in Italy, and when Mussolini formed an alliance with Hitler, he, too, began a purge against Jews. Maja's husband, Paul Winteler, was not a Jew, but he was sick at the time she planned to emigrate, so he stayed behind. She planned on staying in the United States only until hostilities were over, but she eventually became ill and wasn't able to return. As a result, she never saw her husband, or Europe, again.

With Maja's arrival, the household now consisted of three women—Dukas, Maja, and Margot—and Einstein, all living in the rather modest house at 112 Mercer Street in Princeton. Dukas had the responsibility for looking after Einstein's needs; she looked after his correspondence, shielded him from the public, and acted as his housekeeper. Einstein got a lot of hate mail because of his pacifist view and because he was later considered to be the "superfather" of the atomic bomb. Dukas was diligent in shielding him from it. She was firmly dedicated to Einstein and eventually became indispensable to him. His real affection, however, was reserved for Maja and Margot. Over the years both became very close to him. Margot worried about him whenever he was out of her sight. She was described as sweet, but a little lost in both Einstein's and Dukas's world. She enjoyed art and sculpturing, and Einstein encouraged her.

The relationship between Maja and Einstein, according to Einstein's friends, was particularly special. They had had their differences when they were young, but now the two of them were inseparable. Einstein was much more affectionate to her than he was to either of his wives. And strangely, as they aged, they began to look more and more like one another.

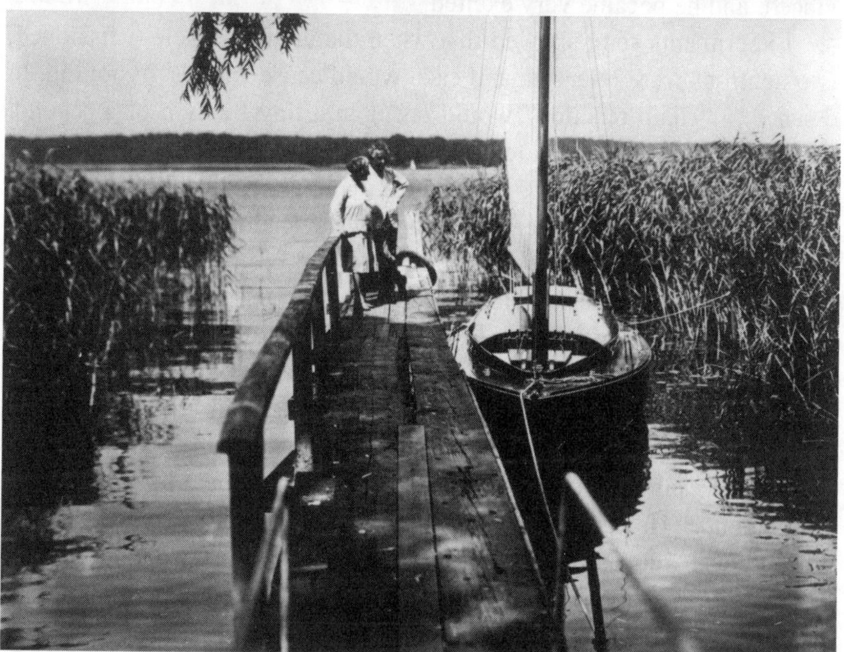

Fig. 27: Einstein and his sailboat

PRELUDE TO THE ATOMIC BOMB

Near the end of 1938 two German scientists, Otto Hahn and Fritz Strassman of the Kaiser Wilhelm Institute in Berlin, made a startling discovery. Upon bombarding uranium nuclei with neutrons, they discovered that there was barium in the decay products. This was particularly surprising since barium is only half as heavy as uranium. They were confused by the result and had no idea what had happened, but they knew it was important. Hahn got in touch with a former colleague, Lise Meitner, who had escaped to Holland, and she and her nephew, Otto Frisch, looked into the problem. After considerable discussion, they finally realized that the presence of barium could only be explained by a breaking up or "fissioning" of the uranium nucleus. It was hard to believe, but they were soon convinced that it was the only answer. Fritsch told Niels Bohr about it, and Bohr, who was just leaving for the United States, was quickly convinced, and he became very excited.

Experiments soon showed that when the uranium nucleus fissioned, two neutrons were released, and each would cause another fissioning. In essence, a "chain reaction" would occur in a tiny fraction of a second, releasing a tremendous amount of energy. Bohr realized the potential for a super bomb, and he conveyed the information to Enrico Fermi when he got to the United States. Fermi was one of the leading nuclear physicists in the world and had barely missed discovering fission himself.

Leo Szilard, who had fled Europe several years earlier, also soon heard of it and almost went into a panic. He had already considered the possibility of a chain reaction involving neutrons, and had actually taken a patent out on it. Realizing the importance of uranium to the discovery, Szilard began to worry that Germany might get its hands on the huge uranium reserves in the Belgian Congo, and he felt he should try to do something about it. He had worked with Einstein earlier and knew that he was a friend of the queen of Belgium. Along with physicist, Eugene Wigner, he therefore drove to Einstein's Long Island summer home to talk to him.

Einstein was not an expert in nuclear physics, and had heard nothing about a chain reaction involving uranium nuclei, but he quickly understood the implications of the breakthrough. He was hesitant about writing to the queen of Belgium, but offered to write to a Belgian cabinet minister whom he knew. Szilard told Einstein he would compose the letter and have him look it over, and if he was satisfied, he hoped he would sign it. Szilard soon decided that a letter to President Franklin Roosevelt might also be useful, but he would have to find someone to deliver it to him personally. He remembered Alexander Sachs, a vice president of the Lehman Corporation, and got in touch with him. And, indeed, Sachs agreed to hand-deliver the letter to Roosevelt. Szilard ended up composing two letters, a short one and a long one, and he returned with them to Einstein for his signature. This time Edward Teller went with him. Einstein signed both letters, but in the end it was the longer letter that Sachs carried to Roosevelt.[5]

As Sachs was trying to deliver the letter, Germany invaded Poland and World War II began. For the next several weeks, Roosevelt was therefore extremely busy, and Sachs couldn't get an appointment with him. Finally, however, many weeks later, Roosevelt offered to meet him, and to Sachs's relief, Roosevelt took the threat that Germany might build an atomic bomb seriously. He agreed to set up a committee to look into the matter. The committee met, and after some discussion, allotted a paltry $6,000 for research into the feasibility of an atomic bomb project. To Szilard, this was an insult; he knew that a much larger amount would be needed. As many more months passed, and everyone seemed to be dragging their feet, Szilard got angry and went to Einstein with a second letter. Einstein signed it, and it was sent on March 7, 1940. But, as it turned out, it would still be well over a year before things really got moving.

EINSTEIN GETS U.S. CITIZENSHIP AND THE UNITED STATES ENTERS THE WAR

By now Einstein had decided that he was going to stay in the United States, and he thought he should become a citizen. He had always cherished his Swiss citizenship and didn't want to lose it. Fortunately, there was no stipulation that he had to. Since he had to apply from abroad, he, along with Margot and Dukas, made a trip to Bermuda in May 1935. They took their oath five years later on October 30, 1940.

It was about this time that Einstein got a letter from his teen sweetheart, Marie Winteler. She was caught in the economic upheaval in Europe and was having difficulty getting enough food. She asked Einstein for a loan of 100 francs (a rather small amount), reminding him that her mother had been particularly kind to him. This was, of course, true; Einstein had regarded Pauline Winteler as his second mother. In a second letter about three months later, she asked Einstein to help her and her son emigrate to the United States. There is no indication she ever received any help, and she did not come to the United States. It seems that Dukas, who handled all of Einstein's correspondence, did not pass the letters on to Einstein. Einstein had such a soft spot for anyone with a problem, he certainly would have helped her if he had known. Dukas shielded him from anything that she thought might affect him emotionally, but she may have gone a little too far in shielding him from Marie's letters. They were found in his correspondence after his death.

Einstein was anxious for the United States to get into the war. He knew it would make a big difference, and he was discouraged that many people were trying to keep America out of the war. By now Hitler had expanded across Europe, conquering country after country; indeed, most of Europe was now under the Nazi flag. But throughout 1940 and most of 1941, America remained aloof. Roosevelt felt that the United States should be in the war, but he had a lot of opposition. On December 7, 1941, Japan attacked Pearl Harbor with devastating force, and the United States had no alternative. It declared war on both Japan and Germany.

This may have been good news to Einstein, but it was a serious and grave commitment. There was, however, some good news in 1941. The atomic bomb project, or "Manhattan Project," as it was called, finally got underway.[6] The first step was to show that a slowed-down version of the fission chain reaction was possible and could be controlled. Enrico Fermi and his team accomplished this at the University of Chicago on December 2, 1942. Then it was full-steam ahead toward an atomic bomb, but as it turned out, success was not achieved until 1945. Einstein did not know of the details of the project, but he was informed by his friend and colleague Otto Stern in the fall of 1944 that it looked like a bomb would be built.

Einstein had to have been shocked. He knew it would eventually be possible, but he was sure it wouldn't come in his lifetime. When he heard of its likely success, he worried about the implications, and his strong pacifist feelings began to reemerge. He hated to see it used. Despite his reservations, though, Einstein was fervently patriotic and wanted to do his part in the war effort. He therefore became a navy consultant in 1943, making calculations related to various types of explosions. He made a much more important contribution to the war effort in 1944. A committee had been set up for the auctioning of rare books and manuscripts, with the proceeds going to help finance the war. They asked him for his original 1905 manuscript on special relativity. He had discarded it years earlier, but offered to rewrite it. At auction it fetched 6 million dollars, which went directly into the war effort. Another of his manuscripts fetched 5 million dollars.

There is a certain irony in all this. As hard as it is to believe, FBI director J. Edgar Hoover was determined to prove that Einstein was a communist and a spy, and for several years he had all his activities carefully monitored.[7] When he couldn't find anything on Einstein, he switched his focus to his secretary, Helen Dukas (who was about as big a danger to our country as a lap poodle). At one point he considered bugging Einstein's telephone, but decided against it because of possible embarrassment to the FBI if the bug were discovered.

In May 1945 the war in Europe ended, and Einstein felt considerable relief, but continued to worry about the atomic bomb. Through Otto Stern, he was kept abreast of some of the developments on the atomic bomb, and was concerned about its use. Szilard was equally worried, and he was also worried about what its role would be after the war. He therefore approached Einstein to sign another letter to Roosevelt, which he did. But before the letter got to Roosevelt, he died. And within a short time, under orders from President Harry S. Truman, atomic bombs were dropped on Hiroshima and Nagasaki.

Einstein heard the news over the radio while he was vacationing at

Saranac Lake. According to Dukas, his only comment was, "Alas, Oh my God!"[8] He made no public statement. Einstein had visited Japan in the early 1920s, and had been well received. Despite the fact that they were our enemies, he did not feel a great deal of animosity toward them, and after the bombing he expressed considerable compassion for them. As far as the Germans were concerned, however, he felt nothing but anger. He blamed them for Hitler's death camps and world aggresssion until his dying day. When Arnold Sommerfeld wrote to him after the war, asking him to let bygones be bygones and accept membership in the Prussian Academy again, he refused. He wanted nothing more to do with the Germans, but he did exempt most of his German friends.

Einstein's contribution to the atomic bomb project—his formula $E = mc^2$, which is at the basis of it, and his letter to Roosevelt initiating it—soon became well known, and as a result, he was sometimes referred to as the "superfather" of the bomb. (Robert Oppenheimer, the director of the "Manhattan Project" at Los Alamos, New Mexico, was referred to as the "father.") Einstein didn't like this and tried to distance himself as much as possible from the bomb. He worried about how the bomb would be used in the new "atomic age." To him, the only solution was a "world government." This world government would be an international organization with powers to control war and the use of the atomic bomb. Einstein called on others to help him in his effort to get someone to organize a world government, but nothing ever came of his effort. Part of his wish was granted, however, when the "United Nations" was formed after the war.

Einstein's guilt over his role in the building of the atomic bomb was so great that at one point he said, "Had I known that the Germans would not succeed in getting the atomic bomb, I never would have supported it."[9] He knew, however, that even without his help, things would have proceeded in about the same way. His initial enthusiasm for the project was, indeed, a result of his worry that the Germans would get the bomb first. After all, fission was discovered by German scientists, Hahn and Strassman, and with the delays in the American project, it appeared as if the Germans might have a lead. As it turned out, they never did. Germany's leading scientist, Werner Heisenberg, had, indeed, talked to Albert

Speer, who was the Nazi minister of armaments, about the possibility of an atomic bomb, and Speer, in turn, talked to Hitler. But Heisenberg scared them by saying it might be very difficult to contain the explosion. Furthermore, he was sure it would take at least three years to complete, assuming there was ample support for the project. Hitler was sure the war would be over by then; nevertheless, he did initiate a halfhearted project, but it never got very far.

DETERIORATING HEALTH

After the war Einstein's health began to deteriorate, and he was never completely healthy again. In addition, Maja had a stroke in 1946, and her health was soon even worse than his. Einstein's affection and worry over her health amazed his friends. She was bedridden after the stroke, and Einstein sat at her bedside for hours, reading and talking to her.

Despite his health, Einstein continued his work on pacifism. He urged the United Nations to form a world government, and in December of 1946 he made a speech that was broadcast by NBC, calling again for a world government. As great as his passion for peace was, his greatest passion was always for his equations and his theories. He eventually returned to his nonsymmetric unified field theory with the hopes that it would finally be the answer. But, as earlier, things didn't gel, and he had to abandon it a second time.

In 1948 Einstein began to have severe abdominal pains, and a cyst was diagnosed by his doctors. In December 1948 he was operated on at the Jewish hospital in Boston. The surgeon found several adhesions of the intestine, but he also found a huge aneurysm in a main artery going to the heart. The artery had expanded at one point to the size of a grapefruit. The surgeon knew it was a significant threat, but it was also dangerous to operate on it. Checking the walls of the aneurysm, he found that they were still relatively thick, and it wasn't likely to rupture, so he left it. Eighteen months later it was checked again, and it had enlarged considerably.

Fig. 28: Einstein in his study at Princeton

MILEVA DIES

Although Einstein did not correspond with Eduard over the years, he continued to worry about his welfare. Eduard, who had been in and out of sanitariums, was now thirty-seven years old, and considerably overweight. Mileva had devoted her life to him, visiting him frequently in the sanitariums and taking him home with her whenever she could. But he was unpredictable, and she sometimes had a difficult time controlling him. Einstein continued sending money for him, but his sanitarium expenses were extremely high. Mileva had bought three houses with Einstein's Nobel Prize money, one of which she lived in. The other two were used for rental income. With the mounting expenses for Eduard, she was eventually forced to sell two of the houses, and there was some fear that she might lose the one in which she lived. She therefore decided to sign it over to Einstein, so she wouldn't lose it.

In 1947 Mileva slipped on some ice and broke her leg. From then on, her health deteriorated rapidly. Einstein worried about what would happen to Eduard if she died, so he decided to sell the house and put the money into a trust for him. A deal was struck in which Mileva would not be forced to move and could live in the house as long as she wished. When Mileva heard that he was selling the house, she was devastated; she would now have lost all three of her houses, and she refused to write to Einstein or reply to his letters. The money from the sale of the house was delivered to her, and she was supposed to forward it to Einstein, who was going to put it in a trust for Eduard. Einstein waited and waited, but heard nothing from her. Finally, he got somebody in Zurich to check on her. He soon learned that she had had a severe stroke that had paralyzed one side of her body, and she was in the hospital. The stroke apparently occurred during one of Eduard's visits home, when she was trying to restrain him.

Mileva died on August 4, 1948.[10] All in all, she led a rather bleak and

Fig. 29: Another view of Einstein in his study at Princeton

tragic life. She rarely spoke to others about Einstein or their life together, but she did have tremendous respect for him and his work, despite their differences. The money from the sale of the house—85,000 francs—was found under the mattress of her bed.

MAJA DIES

During 1949 Einstein recuperated in Florida. He walked the sandy beaches and sailed whenever he could. His love for sailing equaled his love for music, and although he was now to weak too play his violin, he could still manage to sail occasionally. He was at peace with himself and the world when he was out in his sailboat. The only passion in his life that was greater was his love for his equations and his obsession for a unified field theory.

Maja was now completely bedridden. She had hoped to eventually get back to Europe, but it never happened. In 1951 she fell and broke her arm; then as she lay on her bed, she developed pneumonia. She could hardly talk as she lay near death, but Einstein stayed at her side. When she died, he was devastated. He took her death much harder than that of either of his wives. She died on June 25, 1951. "Now I miss her more than can be imagined," he said.[11]

THE FORMATION OF THE STATE OF ISRAEL

Einstein was a strong supporter of the state of Israel, and when Chaim Weizman, its first president, died on November 9, 1952, Einstein got an honor that deeply touched him. Abba Eban, the Israeli ambassador to the United States, asked him if he would accept the presidency of Israel. Einstein had to have been shocked when the offer came, and he knew he couldn't have done it. He didn't have the expertise to guide a newly forming country, and he knew it. Surprisingly, they told him he could continue with his research, even if he took the job. With an offer like this, it seems as if they were looking to him more as a figurehead than anything. Besides his lack of expertise, however, his health was continuing to deteriorate, and he was reluctant to leave the United States. He declined the offer. Einstein was also a strong supporter of the Hebrew University of Jerusalem. Indeed, when he died, he left all his papers to the Hebrew University.

In 1954 Einstein developed hemolytic anemia, and his aneurysm was now considerably larger, and continuing to increase in size. He knew his days were numbered.

Epilogue

Einstein celebrated his seventy-sixth birthday on March 14, 1955. His friends wanted a large celebration, but Einstein was against it, so only a few close friends celebrated it with him. A week later he received news of the death of his longtime friend, Michele Besso. Besso had helped him with his special theory of relativity, and had frequently been a go-between for him and Mileva. He had been a close friend for many years, and Einstein took his death hard. He wrote a moving letter to the Besso family, which included the statement, "What I most admired about him as a human being is that he managed to live so many years not only in peace but also in lasting harmony with a woman—an undertaking in which I twice failed rather disgracefully."[1]

On April 11 Einstein signed an appeal drawn up by British philosopher Bertrand Russell against the arms race. As Einstein had feared, Russia had the atomic bomb, and both Russia and the United States were stockpiling large numbers in the event of a nuclear war. Einstein abhorred the action. Celebrations were also being planned for the seventh anniversary of the founding of the state of Israel, and Einstein had been asked to give a speech. Too weak to travel to Israel, he had offered to make one

Fig. 30: Einstein in old age

that could be recorded, then broadcast over radio in Israel. He was still working on the speech on April 13 when he got pains in his abdomen. Such pains were not new to him. His aneurysm had grown for several years, and he knew it was still increasing in size. Furthermore, it was possible that it could burst anytime. These pains, however, soon became much more intense than any he had experienced previously. Dukas looked after him at first, but it was soon obvious they had to call the doctor. His doctor feared that the aneurysm had been perforated and urged him to go to the Princeton Hospital, but Einstein was reluctant.

As friends gathered around his bedside, his doctors continued to worry. There was a good chance the aneurysm could burst, and he would then have no chance. Encouraged by his friends, Einstein finally agreed to go to the hospital. By April 13 he was considerably worse, and there was now talk of surgery. Einstein was against it; he didn't want his life prolonged artificially. He didn't want to live it out as a vegetable. "I would like to go when I want to go," he said.[2] He wanted to die gracefully, without a lot of fuss.

His son Hans arrived from California. Other friends—Walter Bucky, Otto Nathan, and several doctor friends—arrived from New York. Everyone urged him to undergo the surgery. Several of them looked to Hans to try to persuade him. Einstein was now on intravenous medication, and in considerable pain, but he continued to refuse surgery.

Then, on Sunday, April 17, he felt much better. He was still working on the speech that would be broadcast in Israel, and he asked for the draft he had written. He also spent some time working on his latest unified field theory, confiding to Otto Nathan that he was confident that he was very close to success. When he went to sleep that night, he did not seem to be in imminent danger, but about 1:00 A.M. Monday morning, he woke in tremendous pain. The aneurysm had burst. The nurse rushed in as Einstein said something in German, but she did not understand it. Within minutes he was dead.

When doctors opened him up to look at the aneurysm, they saw that he was in no shape for surgery, and even if he had undergone it, he likely would have died. Einstein wanted no fuss made over him; he wanted no funeral, no gravestone. He wanted to be cremated. A simple ceremony was held at the crematorium, with Otto Nathan saying a few last words. His ashes were then spread in the Delaware River.

The only thing not cremated was his brain. Dr. Thomas Harvey, who performed the autopsy, got Hans's permission to keep the brain.[3] Dukas and Nathan were not pleased, but there was little they could do. Harvey kept it in formaldehyde for years, giving out samples to various individuals and universities for study, and over the years it has been extensively studied. Considering the contributions he made to science, and his obvious genius, one would think that it would look quite different from other brains. As it has turned out, there is some controversy. Dr. Marion Diamond of the University of California at Berkeley studied thin slices of it under the microscope and reported that it was distinctly different. She says she found a much higher number of glial cells (cells that provide the structure of the brain) in the section of his brain that deals with mathematics, as compared to a normal brain. Dr. Lucy Rorke of Philadelphia, however, does not agree. She also studied several thin slices, and thought they looked normal. She did agree, however, that the brain was well preserved and there was no sign of Alzheimer's disease. Others have found that the Sylvan fissure (involved in hearing and speech development) on the surface of Einstein's brain was different from most others, and his parietal lobes were larger than normal. This is the section of the brain in which mathematical intuition is located.

Einstein was, without a doubt, one of the greatest scientists who ever lived, perhaps the greatest. *Time* magazine selected him in the year 2000 as the "Person of the Century," and his legacy is, indeed, incredible. He changed our view of space and time with his special theory of relativity in 1905. He showed us that objects shorten in the direction of travel when they are moving at high speed relative to us, and he showed us that mass increases with speed relative to us. Furthermore, he demonstrated that time is not absolute, but depends on the motion of an observer relative to us. He also showed that the speed of light is unattainable for matter, and that energy and mass are equivalent. Under the proper conditions, a tremendous amount of energy can be extracted from a small amount of mass. He also gave one of the first proofs of the existence of atoms and molecules. But, strangely, when he received the Nobel Prize in 1922, it was for none of these; it was awarded for showing that light was quantized, in other words, it consisted of particles.

Einstein's greatest contribution, however, came in 1915 when he gave us a new theory of gravity. He showed that acceleration and gravity were related, and that gravity was not an action-at-a-distance force, as had been assumed by Newton, but was "curved space." It was a strange, new concept that many people didn't take seriously at first. But when he was able to make a number of important predictions, and they were borne out, people began to pay attention. His new theory also showed that time was not absolute, but depended on the strength of the gravitational field that the observer was in. The clock of an observer in a strong gravitational field ran slower than the clock of an observer in a weak field.

Einstein applied his new theory to the universe and gave us the first mathematical theory of the universe (cosmology). It overcame many of the problems that Newton had encountered and could not solve. In particular, it explained the expansion of the universe that was discovered in 1929 by Edwin Hubble of Mt. Wilson in California. Furthermore, it predicted bizarre objects that we now call black holes. Black holes are strange objects that pull everything in and allow nothing to escape once it is inside them. Several have now been discovered in the universe, and they have been extensively studied over the past few years.

Einstein spent the last thirty years of his life trying to bring everything together into a theory of everything (TOE), but he did not succeed. Nevertheless, he left humankind many great gifts.

Notes

INTRODUCTION

1. Alice Calaprice, *The Expanded Quotable Einstein* (Princeton: Princeton University Press, 2000), p. 14.

2. Ibid., p. 254.

3. Denis Brian, *Einstein: A Life* (New York: Wiley, 1996), p. 187.

4. Calaprice, *Expanded Quotable Einstein*, p. 157.

5. Ibid., p. 245.

6. Banesh Hoffmann, *Albert Einstein: Creator and Rebel* (New York: Viking, 1972).

7. Jürgen Renn and Robert Schulmann, eds., *Albert Einstein/Mileva Marić: The Love Letters*, trans. Shawn Smith (Princeton: Princeton University Press, 1992), p. 23.

8. Calaprice, *Expanded Quotable Einstein*, p. 27.

9. Ibid., p. 37.

10. Ibid., p. 256.

11. Ibid., p. 245

12. Ibid., p. 165.

13. Ibid., p. 163.

14. Ibid., p. 128.

1. AN EARLY PASSION FOR LEARNING AND MUSIC

1. Maja Winteler-Einstein, "Albert Einstein: A Biographical Sketch," in *The Early Years, 1879–1902*, ed. John Stachel et al., trans. Anna Beck, vol. 1 of *The Collected Papers of Albert Einstein* (Princeton: Princeton University Press, 1987–1998), p. XIX.

2. Ibid., p. XVIII.

3. Ibid.

4. Albrecht Fölsing, *Albert Einstein: A Biography* (New York: Viking, 1997), p. 16.

5. Ibid., p. 22.

6. Maja Winteler-Einstein, "Albert Einstein: A Biographical Sketch," p. XIX.

7. Ibid. p. XX.

8. Albert Einstein, "Autobiographical Notes," in *Albert Einstein: Philosopher-Scientist*, ed. Paul Schilpp (New York: Tudor Publishing, 1951), p. 9.

9. Maja Winteler-Einstein, "Albert Einstein: A Biographical Sketch," p. XXI.

10. Don Howard and John Stachel, eds., *Einstein: The Formative Years, 1879–1909* (Boston: Birkhaüser, 2000), p. 23.

11. Max Talmey, *The Relativity Theory Simplified and the Formation Period of Its Inventor* (New York: Falcon Press, 1932).

12. Fölsing, *Albert Einstein*, p. 26.

13. Alice Calaprice, *The Expanded Quotable Einstein* (Princeton: Princeton University Press, 2000), p. 14.

14. Helen Dukas and Banesh Hoffmann, *Albert Einstein: The Human Side* (Princeton: Princeton University Press, 1979), p. 66.

15. Jagdish Mehra, *Einstein, Hilbert, and the Theory of Gravitation* (Boston: Reidel, 1974), p. 66.

16. Lewis Pyenson, *The Young Einstein: The Advent of Relativity* (Bristol, U.K.: Adam Hilger, 1985), p. 35.

17. Maja Winteler-Einstein, "Albert Einstein: A Biographical Sketch," p. XXI.

2. LEAVING MUNICH

1. Maja Winteler-Einstein, "Albert Einstein: A Biographical Sketch," in *The Early Years, 1879–1902*, ed. John Stachel et al., trans. Anna Beck, vol. 1 of *The Collected Papers of Albert Einstein* (Princeton: Princeton University Press, 1987–1998), p. XXI.

2. Albrecht Fölsing, *Albert Einstein: A Biography* (New York: Viking, 1997), p. 17.

3. Ibid.

4. Ibid.

5. Banesh Hoffmann, *Albert Einstein: Creator and Rebel* (New York: Viking, 1972), p. 25.

6. Maja Winteler-Einstein, "Albert Einstein: A Biographical Sketch," p. XXI.

7. Ibid.

8. Denis Brian, *Einstein: A Life* (New York: Wiley, 1996), p. 7.

9. Maja Winteler-Einstein, "Albert Einstein: A Biographical Sketch," p. XXII.

10. Fölsing, *Albert Einstein*, p. 34.

11. Maja Winteler-Einstein, "Albert Einstein: A Biographical Sketch," p. XXII.

12. Fölsing, *Albert Einstein*, p. 35.

13. Photographic section, *The Early Years, 1879–1902*, p. 60.

14. "On the Investigation of the State of the Ether," *The Early Years, 1879–1902*, p. 6.

15. Letter to Caesar Koch, *The Early Years, 1879–1902*, p. 6.

3. FIRST LOVE

1. Albin Herzog letter to Gustav Maier, in *The Early Years, 1879–1902*, ed. John Stachel et al., trans. Anna Beck, vol. 1 of *The Collected Papers of Albert Einstein* (Princeton: Princeton University Press, 1987–1998), p. 7.

2. Albrecht Fölsing, *Albert Einstein: A Biography* (New York: Viking, 1997), p. 37.

3. Ibid., p. 39.

4. Gustav Maier letter to Jost Winteler, *The Early Years, 1879–1902*, p. 8.

5. Lewis Pyenson, *The Young Einstein: The Advent of Relativity* (Bristol, U.K.: Adam Hilger, 1985), p. 11.

6. Aargau Kantonsschule Record, *The Early Years, 1879–1902*, p. 8.

7. Inspector's Report on a Music Examination, *The Early Years, 1879–1902*, p. 12.

8. Denis Brian, *Einstein: A Life* (New York: Wiley, 1996), p. 11.

9. Hermann Einstein letter to Jost Winteler, *The Early Years, 1879–1902*, p. 11.

10. Ibid.

11. Letter to Marie Winteler, *The Early Years, 1879–1902*, p. 12.

12. Pauline Einstein letter to Marie Winteler, *The Early Years, 1879–1902*, p. 31.

13. Biographical Notes, *The Early Years, 1879–1902.*

14. Matura Examination: My Future Plans, *The Early Years, 1879–1902*, p. 15.

15. Final Grades, Aargau Kantonsschule, *The Early Years, 1879–1902*, p. 13.

4. STUDENT DAYS AND A NEW LOVE

1. Carl Seelig, *Albert Einstein: A Documentary Biography* (London: Staples Press, 1956), p. 20.

2. Roger Highfield and Paul Carter, *The Private Lives of Albert Einstein* (London: Faber and Faber, 1993), p. 33.

3. Lewis Pyenson, *The Young Einstein: The Advent of Relativity* (Bristol, U.K.: Adam Hilger, 1985), p. 20.

4. Albert Einstein, "Autobiographical Notes," in *Albert Einstein: Philosopher-Scientist,* ed. Paul Schilpp (New York: Tudor, 1951), p. 15.

5. Banesh Hoffmann, *Albert Einstein: Creator and Rebel* (New York: Viking, 1972), p. 85.

6. Ibid.

7. Letter to Mileva Marić, in *The Early Years, 1879–1902*, ed. John Stachel et al., trans. Anna Beck, vol. 1 of *The Collected Papers of Albert Einstein* (Princeton: Princeton University Press, 1987–1998), p. 123.

8. Don Howard and John Stachel, eds., *Einstein: The Formative Years* (Boston: Birkhaüser, 2000), p. 43.

9. Letter from Mileva Marić, *The Early Years, 1879–1902*, p. 34.
10. Letter to Mileva Marić, *The Early Years, 1879–1902*, p. 123.
11. Albrecht Fölsing, *Albert Einstein: A Biography* (New York: Viking, 1997), p. 56.
12. Howard and Stachel, *Einstein: The Formative Years*, p. 64.
13. Denis Brian, *Einstein: A Life* (New York: Wiley, 1996), p. 18.
14. Ibid.
15. Hoffmann, *Albert Einstein: Creator and Rebel*, p. 117.
16. Highfield and Carter, *Private Lives*, p. 65.
17. Howard and Stachel, *Einstein: The Formative Years*, p. 67.

5. THE WOMEN IN HIS LIFE

1. Denis Brian, *Einstein: A Life* (New York: Wiley, 1996), p. 17.
2. Barry Parker, *Einstein's Brainchild* (Amherst, N.Y.: Prometheus Books, 2000), p. 33
3. Letter to Mileva Marić, in *The Early Years, 1879–1902*, ed. John Stachel et al., trans. Anna Beck, vol. 1 of *The Collected Papers of Albert Einstein* (Princeton: Princeton University Press, 1987–1998), p. 129.
4. Ibid.
5. Verse in the Album of Anna Schmid, *The Early Years, 1879–1902*, p. 128.
6. Letter to Mileva Marić, *The Early Years, 1879–1902*, p. 130.
7. Letter to Julia Niggli, *The Early Years, 1879–1902*, p. 129.
8. Jürgen Renn and Robert Schulmann, eds., *Einstein/Marić: The Love Letters*, trans. Shawn Smith (Princeton: Princeton University Press, 1992), p. 12.
9. Ibid., p. 13.
10. Ibid., p. 16.
11. See *The Love Letters.*
12. Dennis Overbye, *Einstein in Love* (New York: Penguin, 2001), p. 54.
13. Ibid.
14. Albrecht Fölsing, *Albert Einstein: A Biography* (New York: Viking, 1997), p. 56.
15. Private communication to Carl Seelig, May 15, 1952, at ETH.
16. Brian, *Einstein: A Life*, p. 25.
17. Ibid.
18. Ibid., p. 26.

6. FAMILY TIES

1. Letter to Mileva Marić, in *The Early Years, 1879–1902*, ed. John Stachel et al., trans. Anna Beck, vol. 1 of *The Collected Papers of Albert Einstein* (Princeton: Princeton University Press, 1987–1998), p. 142.

2. Letter to Mileva Marić, *The Early Years, 1879–1902*, p. 147.

3. Letter to Mileva Marić, *The Early Years, 1879–1902*, p. 159.

4. Einstein letter to Marić, in *Einstein/Marić: The Love Letters*, ed. Jürgen Renn and Robert Schulmann, trans. Shawn Smith (Princeton: Princeton University Press, 1992), p. 21.

5. Albrecht Fölsing, *Albert Einstein: A Biography* (New York: Viking, 1997), p. 82.

6. Hermann Einstein letter to Wilhelm Ostwald, *The Early Years, 1879–1902*, p. 164.

7. Mileva Marić letter to Helene Savić, *The Early Years, 1879–1902*, p. 183.

8. Einstein letter to Marcel Grossmann, *The Early Years, 1879–1902*, p. 290.

9. Letter from Mileva Marić, *The Early Years, 1879–1902*, p. 168.

10. Einstein letter to Marić, *Love Letters*, p. 46.

11. Mileva Marić to Helene Savić, *The Early Years, 1879–1902*, p. 172.

12. Ibid.

7. MORE DIFFICULTIES

1. Einstein letter to Marić, in *Einstein/Marić: The Love Letters*, ed. Jürgen Renn and Robert Schulmann, trans. Shawn Smith (Princeton: Princeton University Press, 1992), p. 54.

2. Ibid.

3. Ibid., p. 55.

4. Don Howard and John Stachel, eds., *Einstein: The Formative Years, 1879–1909* (Boston: Birkhaüser, 2000), p. 107.

5. Einstein letter to Marić, *Love Letters*, p. 56.

6. Ibid.

7. Marić letter to Einstein, *Love Letters*, p. 57.

8. Albrecht Fölsing, *Albert Einstein: A Biography* (New York: Viking, 1997), p. 86.

9. Roger Highfield and Paul Carter, *The Private Lives of Albert Einstein* (London: Faber and Faber, 1993), p. 81.

10. Denis Brian, *Einstein: A Life* (New York: Wiley, 1996), p. 42.

11. Mileva Marić letter to Helene Savić, in *The Early Years, 1879–1902*, ed. John Strachel et al., trans. Anna Beck, vol. 1 of *The Collected Papers of Albert Einstein* (Princeton: Princeton University Press, 1987–1998), p. 183.

12. Marić letter to Einstein, *Love Letters*, p. 62.

13. Ibid., p. 63.

14. Ibid.

15. Einstein letter to Marić, *Love Letters*, p. 69.

16. Ibid., p. 66.

17. Ibid., p. 71.

18. Ibid., p. 70.

19. Ibid.

20. Highfield and Carter, *Private Lives*, p. 90.

21. Einstein letter to Marić, *The Early Years, 1879–1902*, p. 72.

22. Fölsing, *Albert Einstein*, p. 96.

23. Ibid., p. 107.

8. GAINING NEW INSIGHTS

1. Albrecht Fölsing, *Albert Einstein: A Biography* (New York: Viking, 1997), p. 96.

2. Letter to Mileva Marić, in *The Early Years, 1879–1902*, ed. John Srachel et al., trans. Anna Beck, vol. 1 of *The Collected Papers of Albert Einstein* (Princeton: Princeton University Press, 1987–1998), p. 193.

3. Roger Highfield and Paul Carter, *The Private Lives of Albert Einstein* (London: Faber and Faber, 1993), p. 96.

4. Maurice Solovine, *Letters to Solovine* (Paris: Gautier Villars, 1956), p. 10

5. Ibid.

6. Ibid.

7. Letter from the Swiss Department of Justice, *The Early Years, 1879–1902*, p. 195.

8. Denis Brian, *Einstein: A Life* (New York: Wiley, 1996), p. 52.

9. *The Swiss Years, Writings, 1900–1909*, ed. John Srachel et al., trans. Anna Beck, vol. 2 of *The Collected Papers of Albert Einstein*, p. 56.

10. Ibid., p. 76.

11. Ibid., p. 98.

12. Einstein letter to Marić, *Einstein/Marić: The Love Letters*, ed. Jürgen Renn and Robert Schulmann, trans. Shawn Smith (Princeton: Princeton University Press, 1992), p. 77.

13. Denis Brian, *Einstein: A Life* (New York: Wiley, 1996), p. 53.

9. A PASSION FOR UNDERSTANDING NATURE

1. Albrecht Fölsing, *Albert Einstein: A Biography* (New York: Viking, 1997), p. 107.

2. Roger Highfield and Paul Carter, *The Private Lives of Albert Einstein* (London: Faber and Faber, 1993), p. 101.

3. Fölsing, *Albert Einstein*, p. 112.

4. Ibid.

5. Fölsing, *Albert Einstein*, p. 114.

6. Einstein letter to Einstein-Marić, in *Einstein/Marić: The Love Letters*, ed. Jürgen Renn and Robert Schulmann, trans. Shawn Smith (Princeton: Princeton University Press, 1992), p. 78.

7. G. J. Whitrow, *Einstein: The Man and His Achievements* (New York: Dover, 1967), p. 19.

8. Ibid.

9. John Stachel, *Einstein from B to Z* (Boston: Birkhaüser, 2002), p. 39.

10. Dennis Overbye, *Einstein in Love* (New York: Penguin, 2001), p. 181.

11. Albert Einstein, "Autobiographical Notes," in *Albert Einstein: Philosopher-Scientist*, ed. Paul Schilpp (New York: Tudor Publishing, 1951), p. 9.

12. Ibid.

13. Alice Calaprice, *The Expanded Quotable Einstein* (Princeton: Princeton University Press, 2000), p. 295.

14. Ibid., p. 218.

15. Helen Dukas and Banesh Hoffmann, *Albert Einstein: The Human Side* (Princeton: Princeton University Press, 1979), p. 13.

16. Calaprice, *Expanded Quotable Einstein*, p. 248.

17. Ibid., p. 16.

18. Ibid., p. 20.

19. Stachel, *Einstein from B to Z*, p. 89.

20. Ibid., p. 90.
21. Ibid., p. 27.
22. Ibid.
23. John Stachel, *Einstein's Miracle Year* (Princeton: Princeton University Press, 1998), p. 5.

10. THE MIRACLE YEAR: 1905

1. John Stachel, ed., *Einstein's Miracle Year* (Princeton: Princeton University Press, 1998), p. 177.
2. Banesh Hoffmann, *Albert Einstein: Creator and Rebel* (New York: Viking, 1972), p. 55.
3. Albrecht Fölsing, *Albert Einstein: A Biography* (New York: Viking, 1997), p. 124.
4. Ibid., p. 31.
5. Ibid., p. 85.
6. Denis Brian, *Einstein: A Life* (New York: Wiley, 1996), p. 61.
7. Stachel, *Einstein's Miracle Year*, p. 123.
8. Ibid., p. 123.
9. Albert Einstein et. al., *The Principle of Relativity* (New York: Dover, 1923), p. 69.

11. EXTENDING THE THEORY

1. Albrecht Fölsing, *Albert Einstein: A Biography* (New York: Viking, 1997), p. 206.
2. Denis Brian, *Einstein: A Life* (New York: Wiley, 1996), p. 69.
3. Fölsing, *Albert Einstein*, p. 243.
4. Albert Einstein et. al., *The Principle of Relativity* (New York: Dover, 1923), p. 75.
5. Banesh Hoffmann, *Albert Einstein: Creator and Rebel* (New York: Viking, 1972), p. 89.
6. Brian, *Einstein: A Life*, p. 69.
7. Fölsing, *Albert Einstein*, p. 247.
8. Laub letter to Einstein, March 1, 1908, in *The Swiss Years, Correspon-*

dence, 1902–1914*, ed. Martin Klein et al., trans. Anna Beck, vol. 5 of *The Collected Papers of Albert Einstein* (Princeton: Princeton University Press, 1987–1998).

9. Fölsing, *Albert Einstein*, p. 249.

10. Roger Highfield and Paul Carter, *The Private Lives of Albert Einstein* (London: Faber and Faber, 1993), p. 123.

11. Fölsing, *Albert Einstein*, p. 248.

12. Albert Einstein letter to Anna Schmid, May 12, 1909, *The Swiss Years, Correspondence, 1902–1914.*

13. Ronald Clark, *Einstein: The Life and Times* (New York: World Publishing, 1971), p. 124.

14. Dennis Overbye, *Einstein in Love* (New York: Penguin, 2001), p. 181.

15. Ibid.

16. Brian, *Einstein: A Life*, p. 55.

17. Fölsing, *Albert Einstein*, p. 278.

18. Overbye, *Einstein in Love*, p. 196.

19. Fölsing, *Albert Einstein*, p. 278.

12. THE GENERAL THEORY

1. Don Howard and John Stachel, *Einstein and the History of General Relativity* (Boston: Birkhaüser, 1989), p. 48.

2. If a planet Vulcan actually existed, its gravitational field would pull Mercury out of its predicted orbit.

3. Roger Highfield and Paul Carter, *The Private Lives of Albert Einstein* (London: Faber and Faber, 1993), p. 149.

4. Ibid., p. 150.

5. John Stachel, *Einstein from B to Z* (Boston: Birkhaüser, 2002), p. 281.

6. Abraham Pais, *Subtle Is the Lord* (Oxford: Oxford University Press, 1982), p. 210

7. Ibid., p. 216.

8. Dennis Overbye, *Einstein in Love* (New York: Penguin, 2001), p. 263.

9. Denis Brian, *Einstein: A Life* (New York: Wiley, 1996), p. 85.

10. Ibid.

11. Albrecht Fölsing, *Albert Einstein: A Biography* (New York: Viking, 1997), p. 328.

12. Ronald Clark, *Einstein: The Life and Times* (New York: World Publishing, 1971), p. 168.

13. Highsmith and Carter, *Private Lives*, p. 158.

14. Brian, *Einstein: A Life*, p. 87.

15. Overbye, *Einstein in Love*, p. 261.

16. Highfield and Carter, *Private Lives*, p. 155.

17. Overbye, *Einstein in Love*, p. 258.

18. Ibid., p. 261.

19. Ibid., p. 267.

20. Letter to Hans Einstein, in *The Berlin Years, Correspondence, 1914–1918*, ed. Robert Schulmann et al., trans. Ann Hentschel, vol. 8 of *The Collected Papers of Albert Einstein* (Princeton: Princeton University Press, 1987–1998), p. 84.

21. Highfield and Carter, *Private Lives*, p. 181.

22. Overbye, *Einstein in Love*, p. 289.

23. Fölsing, *Albert Einstein*, p. 369.

24. Ibid., p. 373.

25. Ibid., p. 374.

26. Ibid.

27. Ibid.

28. Overbye, *Einstein in Love*, p. 294.

29. Howard and Stachel, *Einstein and the History of General Relativity*, p. 213.

30. Highfield and Carter, *Private Lives*, p. 177.

31. Ibid., p. 178.

13. CONFIRMATION AND A PASSION FOR DETERMINACY

1. Albrecht Fölsing, *Albert Einstein: A Biography* (New York: Viking, 1997), p. 832.

2. Letter to Hans Einstein, in *The Berlin Years, Correspondence, 1914–1918*, ed. Robert Schulmann et al., trans. Ann Hentschel, vol. 8 of *The Collected Papers of Albert Einstein* (Princeton: Princeton University Press, 1987–1998), p. 449.

3. Letter to Mileva Einstein, *The Berlin Years, Correspondence, 1914–1918*, p. 456.

4. Letter from Mileva Einstein, *The Berlin Years, Correspondence, 1914–1918*, p. 465.

5. Letter to Anna Besso, *The Berlin Years, Correspondence, 1914–1918*, p. 489.

6. Letter from Anna Besso, *The Berlin Years, Correspondence, 1914–1918*, p. 490.

7. Ibid.

8. Letter to Michele Besso, *The Berlin Years, Correspondence, 1914–1918*, p. 598.

9. Letter to Mileva Einstein, *The Berlin Years, Correspondence, 1914–1918*, p. 527.

10. Letter to Heinrich Zangger, *The Berlin Years, Correspondence, 1914–1918*, p. 622.

11. Ilse Einstein letter to Georg Nicholai, *The Berlin Years, Correspondence, 1914–1918*, p. 564.

12. Ibid.

13. Alice Calaprice, *The Expanded Quotable Einstein* (Princeton: Princeton University Press, 2000), p. 39.

14. Fölsing, *Albert Einstein*, p. 548.

15. Ibid., p. 440.

16. Ibid., p. 442.

17. Ibid., p. 443.

18. Ibid., p. 444.

19. Ibid., p. 474.

20. Ibid., p. 464.

21. Ibid., p. 467.

22. Abraham Pais, *Subtle Is the Lord* (Oxford: Oxford University Press, 1982), p. 502.

23. Calaprice, *Expanded Quotable Einstein*, p. 260.

24. Fölsing, *Albert Einstein*, p. 556.

25. Calaprice, *Expanded Quotable Einstein*, p. 245.

26. Ibid., p. 76.

27. Denis Brian, *Einstein: A Life* (New York: Wiley, 1996), p. 164.

14. AN OBSESSION FOR UNITY

1. Roger Highfield and Paul Carter, *The Private Lives of Albert Einstein* (London: Faber and Faber, 1993), p. 227.
2. Denis Brian, *Einstein: A Life* (New York: Wiley, 1997), p. 176.
3. Albrecht Fölsing, *Albert Einstein: A Biography* (New York: Viking, 1997), p. 557.
4. Brian, *Einstein: A Life*, p. 174.
5. Highfield and Carter, *Private Lives*, p. 207.
6. Ronald Clark, *Einstein: The Life and Times* (New York: World Publishing, 1971), p. 215.
7. Alice Calaprice, *The Expanded Quotable Einstein* (Princeton: Princeton University Press, 2000), p. 77.
8. Brian, *Einstein: A Life*, p. 251.
9. Ibid., p. 227.
10. Fölsing, *Albert Einstein*, p. 649.
11. Brian, *Einstein: A Life*, p. 238.
12. Fölsing, *Albert Einstein*, p. 654.
13. Brian, *Einstein: A Life*, p. 249.
14. Ibid.
15. Ibid., p. 195.
16. Highfield and Carter, *Private Lives*, p. 215.
17. Brian, *Einstein: A Life*, p. 261.
18. Ibid., p. 291.

15. A DESIRE FOR WORLD PEACE

1. Alice Calaprice, *The Expanded Quotable Einstein* (Princeton: Princeton University Press, 2000), p. 162.
2. Ibid., p. 164.
3. Albrecht Fölsing, *Albert Einstein: A Biography* (New York: Viking, 1997), p. 635.
4. Ibid., p. 675.
5. Ibid., p. 710.
6. Ibid., p. 713.
7. Denis Brian, *Einstein: A Life* (New York: Wiley, 1996), p. 386.

8. Fölsing, *Albert Einstein*, p. 720.

9. Calaprice, *Expanded Quotable Einstein*, p. 178.

10. Roger Highfield and Paul Carter, *The Private Lives of Albert Einstein* (London: Faber and Faber, 1993), p. 253.

11. Fölsing, *Albert Einstein*, p. 731.

EPILOGUE

1. Alice Calaprice, *The Expanded Quotable Einstein* (Princeton: Princeton University Press, 2000), p. 75.

2. Albrecht Fölsing, *Albert Einstein: A Biography* (New York: Viking, 1997), p. 740.

3. Denis Brian, *Einstein: A Life* (New York: Wiley, 1996), p. 437.

Glossary

ABSOLUTE MOTION. Motion that is the same regardless of the system it is measured in.

ABSOLUTE TIME. A universal time that is the same for all observers in the universe, independent of their motion.

ACCELERATION. The rate of change of velocity.

AXIOM. A self-evident truth.

BROWNIAN MOTION. The zigzag motion of tiny grains such as pollen on a fluid. Can be observed with a microscope.

CALCULUS. An advanced branch of mathematics that deals with tiny changes in time and other variables.

CAPILLARITY. The rise of water or other fluid in a narrow or very fine tube.

CATHODE RAYS. A negative current of particles. Electrons.

CAUSALITY. The principle that says cause must come before effect.

CHAIN REACTION. A reaction in which the fission of one atomic nucleus gives off enough neutrons to cause the fission of two or more other nuclei.

CLASSICAL THEORY. Any nonquantum theory. Newton's theory, Maxwell's electromagnetic theory, and relativity theory are examples.

COSMOLOGICAL CONSTANT. A constant Einstein added to his equations of general relativity to stabilize the universe.

COSMOLOGY. The study of the structure and evolution of the universe.

COVARIANCE. Implies that the form of the equation remains the same in any transformation.

DENSITY. Mass per unit volume.

DETERMINISM. The idea that everything in the universe is determined by specific laws.

DIFFERENTIAL EQUATION. An equation in calculus. Can give the time evolution of a system.

DYNAMICS. The study of forces and the motions they cause.

ELECTRIC FIELD. The field around an electric charge.

ELECTRODYNAMICS. The study of the interactions of charged particles.

ELECTROMAGNETIC FORCE. The force that arises between two charged particles.

ELECTROMAGNETIC WAVE. A wave given off by oscillating electric charges.

ELECTRON. The basic particle of electric current. Also a component of the atom. Negatively charged.

ELEMENTARY PARTICLE. A fundamental particle of nature. Particle from which all other structures or particles are built.

ETHER. A hypothetical substance believed at one time to permeate the universe. Needed to propagate waves.

EUCLIDEAN GEOMETRY. Geometry devised by Euclid, based on axioms and theorems.

FISSION. To break in half. The uranium nucleus fissions when it becomes unstable.

FREQUENCY. The number of vibrations per second.

GEODESIC. The shortest distance between two points. Can also be the longest.

HEAT CONDUCTION. The transfer of heat by molecular vibrations.

HELIOSTAT. A system of mirrors that directs the Sun's image to a photographic plate.

HYDRODYNAMICS. The study of the dynamics of fluid flow.

INDETERMINACY. The inability to determine something exactly.

INERTIA. The resistance to change in motion.

INTERFEROMETER. A device that combines light sources and gives an interference pattern.

ISLAND UNIVERSE. Refers to a large group of stars. Older term for galaxy.

KINETIC ENERGY. The energy of motion.

KINETIC THEORY OF GASES. Theory of gases that assumes gases are composed of molecules that are in motion and have energy.

LAWS OF MOTION. The basic laws devised by Newton explaining how and why object move.

MAGNETIC FIELD. Field of a magnet.

MASS. A measure of the amount of matter in a body.

MATRIX. An array of numbers used in mathematics.

MECHANICS. The study of the motion of bodies.

METEOROLOGY. The study of weather.

MICHELSON-MORLEY EXPERIMENT. Experiment of 1887 to determine the motion of Earth through the hypothetical ether.

MOLECULAR FORCE. Force between molecules.

MOMENTUM. Mass multiplied by velocity.

NEWTONIAN MECHANICS. The basic laws of motion postulated by Newton, which explains all dynamical behavior on Earth and in the universe.

NONCOVARIANT THEORY. A theory that is not covariant. In other words the equations change from system to system.

NUCLEAR REACTOR. An installation in which nuclei react with protons, neutrons, and other particles to produce energy.

PACIFIST. A person devoted to peace.

PERTURBATION. A small change.

PHOTOELECTRIC EFFECT. The emission of electrons from a metal when light is shone on it.

PROTON. A particle of light or other electromagnetic radiation.

PRECESSION. A slow change in the orientation of the major axis of an elliptical orbit.

PRINCIPLE OF EQUIVALENCE. Postulates the equivalence of acceleration and gravity.

PRIVATDOZENT. An unpaid (except for student fees) position at a European university. Required for a person trying to become a professor.

PROTON. Heavy particle of the nucleus of an atom. Positively charged.

QUANTUM. A discrete amount of energy that is absorbed or emitted in a particle interaction.

QUANTUM MECHANICS. Theory of atoms and molecules, their structure, and interactions with radiation.

RADIATION. Electromagnetic energy or photons.

RADIOACTIVITY. The emission of various types of particles and radiation from the nucleus.

RELATIVE MOTION. Motion as compared to another (who may also be in motion).

RICCI TENSOR. A tensor devised by Gregorio Ricci that gives a measure of the curvature.

SPACE-TIME. A four-dimensional unification of space and time.

SPECTRUM. Lines seen when light is passed through an instrument that separates out the various frequencies of light.

TENSOR. Component of a complex branch of mathematics called tensor analysis.

THERMOCOUPLE. A device for measuring temperature.

THERMODYNAMICS. The study of the dynamics of heat.

THOUGHT EXPERIMENT. An experiment performed only in the mind.

TIME DILATION. A change in a time interval caused by motion or changed gravitational field of an object.

TRANSFORMATION. A mathematical relation between two systems. A change of coordinates.

UNCERTAINTY PRINCIPLE. Principle that states that there is an uncertainty when we attempt to measure various variables in physics simultaneously.

UNIFIED FIELD THEORY. An attempt to include electromagnetism in general relativity, or more generally, to unify all the fields of nature.

VISCOSITY. A measure of the "stickiness" or internal friction of a fluid.

WAVELENGTH. Distance between equivalent points along a wave.

Bibliography

INTRODUCTION

Brian, Denis. *Einstein: A Life.* New York: Wiley, 1996.

Calaprice, Alice. *The Expanded Quotable Einstein.* Princeton: Princeton University Press, 2000.

Hoffmann, Banesh. *Albert Einstein: Creator and Rebel.* New York: Viking, 1972.

Renn, Jürgen, and Robert Schulmann, eds. *Einstein/Marić: The Love Letters.* Translated by Shawn Smith. Princeton: Princeton University Press, 1992.

1. AN EARLY PASSION FOR LEARNING AND MUSIC

Beck, Anna, trans. *The Early Years, 1879–1902.* Vol. 1 of *The Collected Papers of Albert Einstein.* English translation. Princeton: Princeton University Press, 1987–1998.

Fölsing, Albrecht. *Albert Einstein: A Biography.* New York: Viking, 1997.

Frank, Philipp. *Einstein: His Life and Times.* New York: Knopf, 1972.

Hoffmann, Banesh. *Albert Einstein: Creator and Rebel.* New York: Viking, 1972.

Howard, Don, and John Stachel, eds. *Einstein: The Formative Years.* Boston: Birkhaüser, 2000.

Mehra, Jagdish. *Einstein, Hilbert, and the Theory of Gravitation.* Boston: Reidel, 1974.

Pyenson, Lewis. *The Young Einstein: The Advent of Relativity.* Bristol, U.K.: Adam Hilger, 1985.

Schilpp, Paul, ed. *Albert Einstein: Philosopher-Scientist.* New York: Tudor Publishing, 1951.

Stachel, John, ed. *The Early Years, 1879–1902.* Vol. 1 of *The Collected Papers of Albert Einstein.* Princeton: Princeton University Press, 1987–1998.

Talmey, Max. *The Relativity Theory Simplified and the Formation Period of Its Inventor.* New York: Flacon Press, 1932.

2. LEAVING MUNICH

Beck, Anna, trans. *The Early Years, 1879–1902.* Vol. 1 of *The Collected Papers of Albert Einstein.* English translation. Princeton: Princeton University Press, 1987–1998.

Brian, Denis. *Einstein: A Life.* New York: Wiley, 1996.

Fölsing, Albrecht. *Albert Einstein: A Biography.* New York: Viking, 1997.

Hoffmann, Banesh, *Albert Einstein: Creator and Rebel.* New York: Viking, 1972.

Stachel, John, ed. *The Early Years, 1879–1902.* Vol. 1 of *The Collected Papers of Albert Einstein.* Princeton: Princeton University Press, 1987–1998.

3. FIRST LOVE

Beck, Anna, trans. *The Early Years, 1879–1902.* Vol. 1 of *The Collected Papers of Albert Einstein.* English translation. Princeton: Princeton University Press, 1987–1998.

Brian, Denis. *Einstein: A Life.* New York: Wiley, 1996.

Fölsing, Albrecht. *Albert Einstein: A Biography.* New York: Viking, 1997.

Pyenson, Lewis. *The Young Einstein: The Advent of Relativity.* Bristol, U.K.: Adam Hilger, 1985.

Stachel, John, ed. *The Early Years, 1879–1902.* Vol. 1 of *The Collected Papers of Albert Einstein.* Princeton: Princeton University Press, 1987–1998.

4. STUDENT DAYS AND A NEW LOVE

Beck, Anna, trans. *The Early Years, 1879–1902.* Vol. 1 of *The Collected Papers of Albert Einstein.* English translation. Princeton: Princeton University Press, 1987–1998.

Brian, Denis. *Einstein: A Life.* New York: Wiley, 1996.

Fölsing, Albrecht. *Albert Einstein: A Biography.* New York: Viking, 1997.

Frank, Philipp. *Einstein: His Life and Times.* New York: Knopf, 1972.

Highfield, Roger, and Paul Carter. *The Private Lives of Albert Einstein.* London: Faber and Faber, 1993.

Hoffmann, Banesh, *Albert Einstein: Creator and Rebel.* New York: Viking, 1972.

Howard, Don, and John Stachel, eds. *Einstein: The Formative Years.* Boston: Birkhaüser, 2000.

Pyenson, Lewis. *The Young Einstein: The Advent of Relativity.* Bristol, U.K.: Adam Hilger, 1985.

Schilpp, Paul. *Albert Einstein: Philosopher-Scientist.* New York: Tudor, 1951.

Seeling, Carl. *Albert Einstein: A Documentary Biography.* London: Staples, 1956.

Stachel, John, ed. *The Early Years, 1879–1902.* Vol. 1 of *The Collected Papers of Albert Einstein.* Princeton: Princeton University Press, 1987–1998.

5. THE WOMEN IN HIS LIFE

Beck, Anna, trans. *The Early Years, 1879–1902.* Vol. 1 of *The Collected Papers of Albert Einstein.* English translation. Princeton: Princeton University Press, 1987–1998.

Brian, Denis. *Einstein: A Life.* New York: Wiley, 1996.

Fölsing, Albrecht. *Albert Einstein: A Biography.* New York: Viking, 1997.

Overbye, Dennis. *Einstein in Love.* New York: Penguin, 2001.

Parker, Barry. *Einstein's Brainchild.* Amherst, N.Y.: Prometheus Books, 2000.

Renn, Jürgen, and Robert Schulmann, eds. *Einstein/Marić: The Love Letters.* Translated by Shawn Smith. Princeton: Princeton University Press, 1992.

Stachel, John, ed. *The Early Years, 1879–1902.* Vol. 1 of *The Collected Papers of Albert Einstein.* Princeton: Princeton University Press, 1987–1998.

6. FAMILY TIES

Beck, Anna, trans. *The Early Years, 1879–1902.* Vol. 1 of *The Collected Papers of Albert Einstein.* English translation. Princeton: Princeton University Press, 1987–1998.

Fölsing, Albrecht. *Albert Einstein: A Biography.* New York: Viking, 1997.

Renn, Jürgen, and Robert Schulmann, eds. *Einstein/Marić: The Love Letters.* Translated by Shawn Smith. Princeton: Princeton University Press, 1992.

Stachel, John, ed. *The Early Years, 1879–1902.* Vol. 1 of *The Collected Papers of Albert Einstein.* Princeton: Princeton University Press, 1987–1998.

7. MORE DIFFICULTIES

Brian, Denis. *Einstein: A Life.* New York: Wiley, 1996.

Fölsing, Albrecht. *Albert Einstein: A Biography.* New York: Viking, 1997.

Highfield, Roger, and Paul Carter. *The Private Lives of Albert Einstein.* London: Faber and Faber, 1993.

Howard, Don, and John Stachel, eds. *Einstein: The Formative Years.* Boston: Birkhaüser, 2000.

Renn, Jürgen, and Robert Schulmann, eds. *Einstein/Marić: The Love Letters.* Translated by Shawn Smith. Princeton: Princeton University Press, 1992.

8. GAINING NEW INSIGHTS

Beck, Anna, trans. *The Early Years, 1879–1902.* Vol. 1 of *The Collected Papers of Albert Einstein.* English translation. Princeton: Princeton University Press, 1987–1998.

Brian, Denis. *Einstein: A Life.* New York: Wiley, 1996.

Fölsing, Albrecht. *Albert Einstein: A Biography.* New York: Viking, 1997.

Highfield, Roger, and Paul Carter. *The Private Lives of Albert Einstein.* London: Faber and Faber, 1993.

Renn, Jürgen, and Robert Schulmann, eds. *Einstein/Marić: The Love Letters.* Translated by Shawn Smith. Princeton: Princeton University Press, 1992.

Solovine, Maurice. *Letters to Solovine.* Paris: Gauthier Villars, 1956.

Stachel, John, ed. *The Swiss Years, Writings, 1900–1909.* Vol. 2 of The *Collected Papers of Albert Einstein.* Princeton: Princeton University Press, 1987–1998.

9. A PASSION FOR UNDERSTANDING NATURE

Calaprice, Alice. *The Expanded Quotable Einstein.* Princeton: Princeton University Press, 2000.

Dukas, Helen, and Banesh Hoffmann. *Albert Einstein: The Human Side.* Princeton: Princeton University Press, 1979.

Fölsing, Albrecht. *Albert Einstein: A Biography.* New York: Viking, 1997.

Highfield, Roger, and Paul Carter. *The Private Lives of Albert Einstein.* London: Faber and Faber, 1993.

Overbye, Dennis. *Einstein in Love.* New York: Penguin, 2001.

Renn, Jürgen, and Robert Schulmann, eds. *Einstein/Marić: The Love Letters.* Translated by Shawn Smith. Princeton: Princeton University Press, 1992.

Schilpp, Paul, ed. *Albert Einstein: Philosopher-Scientist.* New York: Tudor Publishing 1951.

Stachel, John. *Einstein from B to Z.* Boston: Birkhaüser, 2002.

———, ed. *Einstein's Miracle Year.* Princeton: Princeton University Press, 1998.

Whitrow, G. J., ed. *Einstein: The Man and His Achievement.* New York: Dover, 1967.

10. THE MIRACLE YEAR: 1905

Brian, Denis. *Einstein: A Life.* New York: Wiley, 1996.

Einstein, Albert, et al. *The Principle of Relativity.* New York: Dover, 1923.

Fölsing, Albrecht. *Albert Einstein: A Biography.* New York: Viking, 1997.

French, A. P. *Einstein: A Centenary Volume.* Cambridge: Harvard University Press, 1979.

Hoffmann, Banesh, *Albert Einstein: Creator and Rebel.* New York: Viking, 1972.

Stachel, John, ed. *Einstein's Miracle Year.* Princeton: Princeton University Press, 1998.

11. EXTENDING THE THEORY

Brian, Denis. *Einstein: A Life.* New York: Wiley, 1996.
Einstein, Albert, et al. *The Principle of Relativity.* New York: Dover, 1923.
Clark, Ronald. *Einstein: The Life and Times.* New York: World Publishing, 1971.
Fölsing, Albrecht. *Albert Einstein: A Biography.* New York: Viking, 1997.
French, A. P. *Einstein: A Centenary Volume.* Cambridge: Harvard University Press, 1979.
Highfield, Roger, and Paul Carter. *The Private Lives of Albert Einstein.* London: Faber and Faber, 1993.
Hoffmann, Banesh. *Albert Einstein: Creator and Rebel.* New York: Viking, 1972.
Klein, Martin, A. J. Kox, and Robert Schulmann, eds. *The Swiss Years, Correspondence, 1902–1914.* Vol. 5 of *The Collected Papers of Albert Einstein.* Princeton: Princeton University Press, 1987–1998.
Overbye, Dennis. *Einstein in Love.* New York: Penguin, 2001.

12. THE GENERAL THEORY

Brian, Denis. *Einstein: A Life.* New York: Wiley, 1996.
Clark, Ronald. *Einstein: The Life and Times.* New York: World Publishing, 1971.
Fölsing, Albrecht. *Albert Einstein: A Biography.* New York: Viking, 1997.
Hentschel, Ann, trans., Robert Schulmann et al., ed. *The Berlin Years, Correspondence, 1914–1918.* Vol. 8 of *The Collected Papers of Albert Einstein.* English translation. Princeton: Princeton University Press, 1987–1998.
Highfield, Roger, and Paul Carter. *The Private Lives of Albert Einstein.* London: Faber and Faber, 1993.
Howard, Don, and John Stachel. *Einstein and the History of General Relativity.* Boston: Birkhaüser, 1989.
Michelmore, Peter. *Einstein: Profile of the Man.* London: Fredrick Muller, 1963.
Overbye, Dennis. *Einstein in Love.* New York: Penguin, 2001.
Pais, Abraham. *Subtle Is the Lord.* Oxford: Oxford University Press, 1982.
Stachel, John. *Einstein from B to Z.* Boston: Birkhaüser, 2002.

13. CONFIRMATION AND A PASSION FOR DETERMINACY

Brian, Denis. *Einstein: A Life.* New York: Wiley, 1996.

Calaprice, Alice. *The Expanded Quotable Einstein.* Princeton: Princeton University Press, 2000.

Fölsing, Albrecht. *Albert Einstein: A Biography.* New York: Viking, 1997.

Hentschel, Ann, trans., Robert Schulmann et al., ed. *The Berlin Years, Correspondence, 1914–1918.* Vol. 8 of *The Collected Papers of Albert Einstein.* English translation. Princeton: Princeton University Press, 1987–1998.

Pais, Abraham. *Subtle Is the Lord.* Oxford: Oxford University Press, 1982.

14. AN OBSESSION FOR UNITY

Brian, Denis. *Einstein: A Life.* New York: Wiley, 1996.

Calaprice, Alice. *The Expanded Quotable Einstein.* Princeton: Princeton University Press, 2000.

Clark, Ronald. *Einstein: The Life and Times.* New York: World Publishing, 1971.

Fölsing, Albrecht. *Albert Einstein: A Biography.* New York: Viking, 1997.

Highfield, Roger, and Paul Carter. *The Private Lives of Albert Einstein.* London: Faber and Faber, 1993.

15. A DESIRE FOR WORLD PEACE

Brian, Denis. *Einstein: A Life.* New York: Wiley, 1996.

Calaprice, Alice. *The Expanded Quotable Einstein.* Princeton: Princeton University Press, 2000.

Fölsing, Albrecht. *Albert Einstein: A Biography.* New York: Viking, 1997.

Highfield, Roger, and Paul Carter. *The Private Lives of Albert Einstein.* London: Faber and Faber, 1993.

Pais, Abraham. *Subtle Is the Lord.* Oxford: Oxford University Press, 1982.

EPILOGUE

Brian, Denis. *Einstein: A Life.* New York: Wiley, 1996.

Calaprice, Alice. *The Expanded Quotable Einstein.* Princeton: Princeton University Press, 2000.

Fölsing, Albrecht. *Albert Einstein: A Biography.* New York: Viking, 1997.

Index

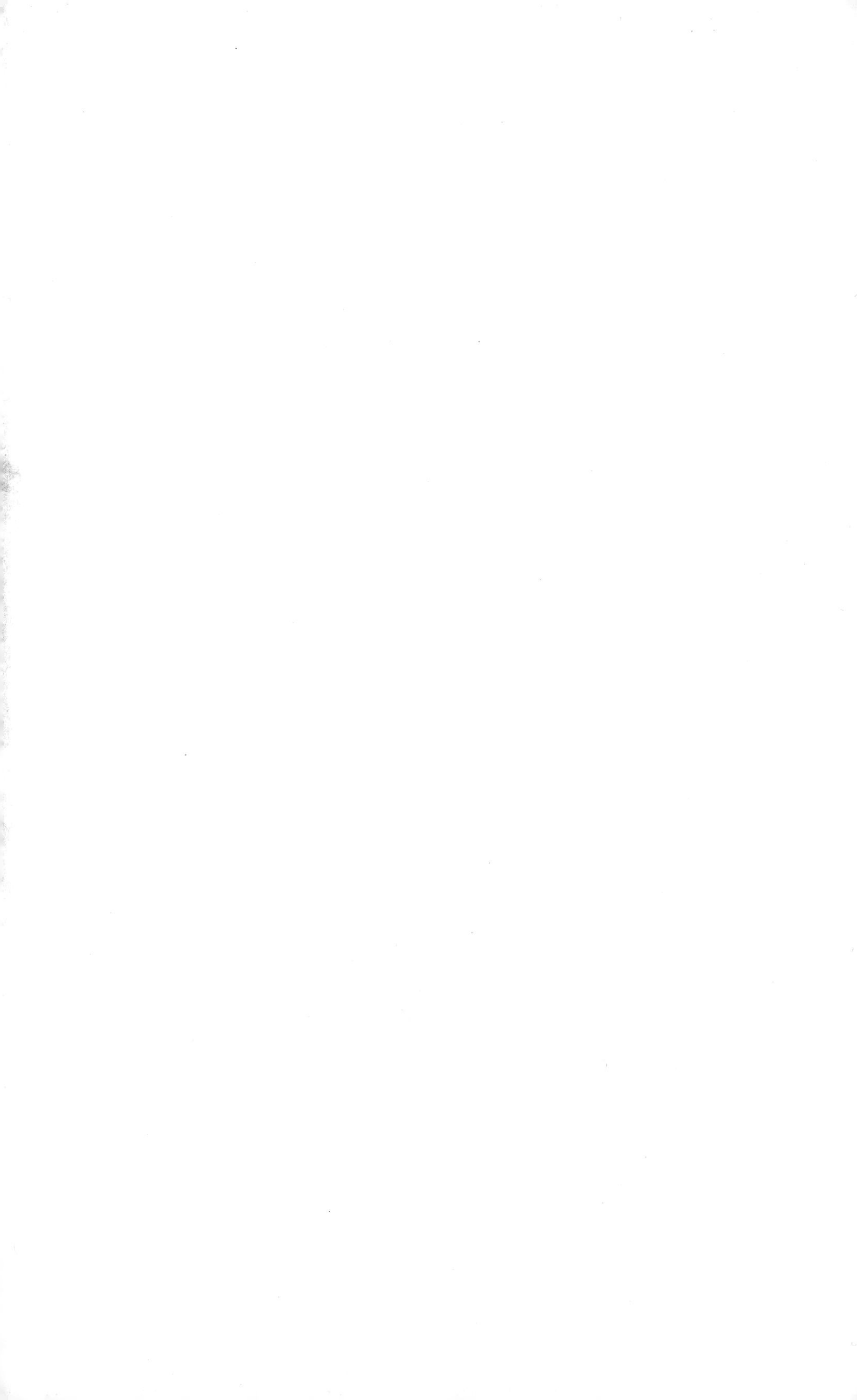

WITHDRAWN